Geothermal Energy Projects

Pergamon Policy Studies on Science and Technology

Carman Obstacles to Mineral Development
Golany Arid Zone Settlement Planning
Goodman/Love Management of Development Projects
Goodman/Love Project Planning and Management
Gould Bureaucratic Corruption and Underdevelopment in the Third World
Stepanek Bangladesh — Equitable Growth?

Related Titles

Goodman/Love Biomass Energy Projects
Goodman/Love Food Projects
Goodman/Love Project Planning and Management
Goodman/Love Small Hydroelectric Projects for Rural Development

Related Journals*

COMPUTERS & OPERATIONS RESEARCH and Their Application to Problems of World Concern : An International Journal.
JOURNAL OF THE OPERATIONAL RESEARCH SOCIETY

*Free specimen copies available upon request.

PERGAMON POLICY STUDIES ON SCIENCE AND TECHNOLOGY

Geothermal Energy Projects
Planning and Management

Edited by
Louis J. Goodman
Ralph N. Love

Published in cooperation with the
East-West Center, Hawaii

Pergamon Press
NEW YORK • OXFORD • TORONTO • SYDNEY • FRANKFURT • PARIS

Pergamon Press Offices:

U.S.A.	Pergamon Press Inc., Maxwell House, Fairview Park, Elmsford, New York 10523, U.S.A.
U.K.	Pergamon Press Ltd., Headington Hill Hall, Oxford OX3 0BW, England
CANADA	Pergamon of Canada, Ltd. Suite 104, 150 Consumers Road, Willowdale, Ontario M2J 1P9, Canada
AUSTRALIA	Pergamon Press (Aust.) Pty. Ltd., P.O. Box 544, Potts Point, NSW 2011, Australia
FRANCE	Pergamon Press SARL, 24 rue des Ecoles, 75240 Paris, Cedex 05, France
FEDERAL REPUBLIC OF GERMANY	Pergamon Press GmbH, Hammerweg 6, Postfach 1305, 6242 Kronberg/Schönberg, Federal Republic of Germany

The East-West Center gratefully extends special acknowledgment to the Exxon Education Foundation for its grant to support the case studies on development projects.

Library of Congress Cataloging in Publication Data

Main entry under title:

Geothermal energy projects.

(Pergamon policy studies)
Bibliography: p.
Includes index.
1. Geothermal engineering. 2. Industrial project management. I. Goodman, Louis J. II. Love, Ralph Ngatata.
TJ2807.G4416 1980 621.44 79-27715
ISBN 0-08-025095-5

Printed in the United States of America

Contents

List of Tables

Table

List of Figures

Figure

List of Symbols

LIST OF SYMBOLS

a	acre
AC	alternating current
amp-hr	ampere-hour
Btu	British thermal unit
bu	bushel
°C	degrees Celsius
cm	centimeter
DC	direct current
°F	degrees Fahrenheit
ft	foot
g	gram
gal	gallon (U.S.)
Gcal	gigacalorie
ha	hectare
hl	hectoliter
hp	horsepower
hr	hour
Hz	Hertz
in.	inch
kcal	kilocalorie
kg	kilogram
km	kilometer
klb	kilopound
kV	kilovolt
kVa	kilovolt ampere
kW	kilowatt
kWe	kilowatt electrical
kWh	kilowatt-hour
l	liter
lb	pound
m	meter
u	micron
mA	milliampere
mi	mile
MHz	megahertz
mph	miles per hour
m/sec	meters per second
MW	megawatt
mV	millivolt
-cm	ohm-centimeter
psig	pounds per square inch, gauge
W	Watt
Wh	watt-hour
Whe	watt-hour electrical

Preface and Acknowledgments

Geothermal energy, now recognized as an important and viable source of alternative energy, is presently being either developed or seriously examined by more than 80 countries interested in the prospect of establishing geothermal industries.

The growth of the geothermal energy industry, as well as its future potential, has undergone a substantial change since the 1950s. From 1958 to 1973, the average growth rate of geothermal power development was not very much more rapid than that for conventional power. In 1973, however, both the quintupling of world oil prices and a restriction of oil supplies dramatically changed the global energy scene. This dual occurrence produced an abrupt change in the economic base of the international energy economy and the cumulative effect of these events was the creation of a sudden worldwide increase of interest in alternative forms of energy – one of which was geothermal. By 1975, the annual growth rates in geothermal plant capacity greatly exceeded the world electricity growth rate.

Attention is focused on the dual problems of: the development of geothermal power as an alternative energy source, and the initiative and management of geothermal development projects in order to exploit this energy. Chapter 1 is a general introduction to the principles of geothermal energy and to the key aspects of geothermal energy development. It also describes the conceptual framework(1) used to analyze the case histories that follow.

Chapters 2, 3, and 4 – case histories of three important geothermal projects in New Zealand, the United States, and the Philippines, respectively – illustrate variations in different socioeconomic settings, in scale and location. Each case is examined as an integrated project cycle, providing detailed analysis and evaluation. With the aid of the single conceptual framework for comparative evaluation, many key common factors in these three unique projects can be compared and contrasted, for example, the organizational complexity of developing geothermal projects.

Chapter 5 summarizes the case findings and provides a comparative evaluation of management techniques in geothermal energy development projects. In addition, it analyzes a number of broad policy and technical issues relating to geothermal development.

This book, the first collection of case histories of geothermal projects to be described from conception to operation within an integrated framework, creates an information base from which policy makers and project managers may draw. The projects described represent the historic foundations of what may prove to be, in the next century, a major rather than an "alternative" energy source for mankind.

The writing of this book would not have been possible without the collaboration and cooperation of many senior scholars and practitioners and their institutions in Asia, the Pacific, and the United States. The general editors wish to convey warmest thanks and deepest appreciation for the many contributions. Space does not permit adequate acknowledgment of each person and institution involved in this international cooperative project effort at the East-West Center's Resource Systems Institute (RSI), but special thanks are due to the authors of the various cases and to their institutions for the splendid cooperation received. In particular, special acknowledgment is given to Richard Bolton, chief geothermal engineer of New Zealand, for his contributions to the book, both as coauthor of the New Zealand case and as advisor for chapters 1 and 5. In addition, warmest thanks are conveyed to John Shupe, dean of the College of Engineering, University of Hawaii and director of the Hawaii Geothermal Project, for his cooperation in facilitating the research necessary for chapter 3 and in reviewing the chapter. A special note of thanks must go to Vicki Nelson for her invaluable advice in working closely with the general editors in the professional editing of the manuscript. Also, thanks are extended to RSI writer/editor Barbara Yount for her advice regarding the overall manuscript and to Jim Mack for his editorial assistance in the preparation of chapter 2.

It is hoped that the impact of this cooperative project will have profound implications in the context of international efforts to improve the planning and management of development projects for all sectors of the economy and society. This should result in more efficient and effective use of critical resources for the mutual benefit of many countries and the well-being of their people.

NOTE

(1) A more detailed explanation of this model, the integrated project planning and management cycle (IPPMC), may be found in our first book on development project cases, Management of Development Projects (Elmsford, N.Y.: Pergamon Press, 1979), and in a second book on the theory and practice of project management, Project Planning and Management: An Integrated Approach (Elmsford, N.Y.: Pergamon Press, 1980).

Geothermal Energy Projects

1 An Introduction to Geothermal Energy

Louis J. Goodman
Ralph N. Love

EARTH HEAT

For centuries, man has been both fascinated by and fearful of the volcanoes, geysers, fumaroles, hot springs, and pools of boiling mud that are the dramatic visual manifestations of the immense reservoir of the earth's heat. The origin of this heat lies beneath the earth's crust where a vast storehouse of immeasurable energy waits to be harnessed by human technological ingenuity.(1) At present, however, less is known about the interior of the earth and its storehouse of energy than is known about the deep oceans and outer space.(2)

What information we have about the inner earth is drawn mostly from indirect knowledge. The structure of the earth consists of five concentric spheres, which are illustrated in figure 1.1. Life on the crust of the earth is dependent upon the atmosphere which surrounds the earth and the heat from beneath the crust. Directly under the crust lies the mantle, toward the center of the earth is the liquid core, and at the earth's center is the inner core. The earth's heat comes from deep within; the nearer one gets to the center, the higher the temperature and the density become.

The emergence of heat in one of its various forms occurs when the crust of the earth is penetrated by the heat source from within. The crust of the earth is made up of six major and a few smaller discrete plates. These are continually in a state of relative motion. Where the plates spread apart, molten rock underlying the crust flows upward; where they move together, one plate goes up and the other goes under and melts into the interior. It is at these points, the junctions of the plates, that heat travels from the hot interior of the earth to the surface, where it may appear as a volcano. Cold surface water may go down to great depths and be heated by high-temperature zones within the crust. The heated water then returns upward by convection and gives rise to spectacular geysers and hot springs.

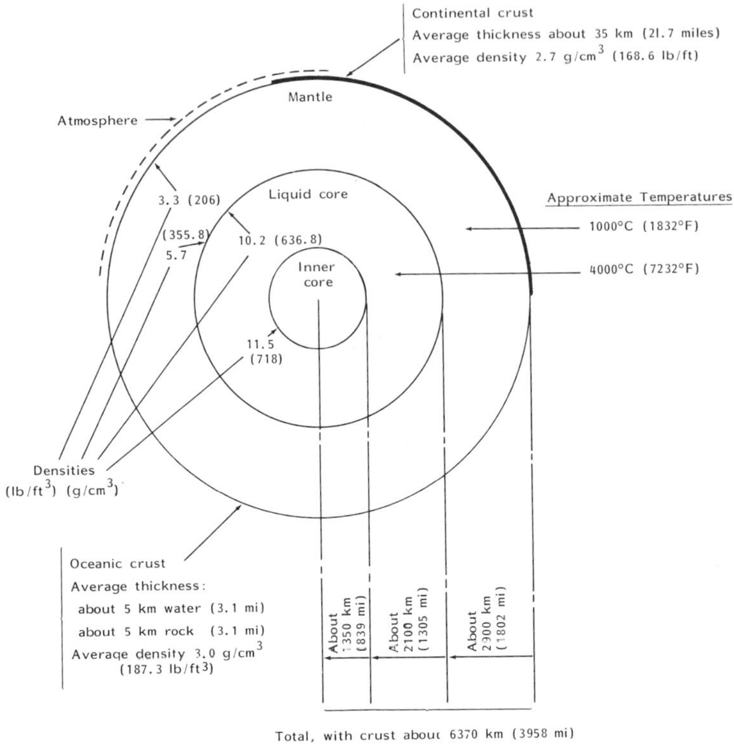

Fig. 1.1. Concentric layers of the earth.
Source: Adapted from H. Christopher H. Armstead, Geothermal Energy (London: E. and F.N. Spon, 1978) p. 19. Reprinted with permission.

The earliest sites where geothermal energy was used by mankind – Italy, New Zealand, Japan, and Iceland – were ones where surface manifestations such as volcanoes, geysers, and hot pools occurred. These early sites, along with other areas in which the earth's heat is now being considered as a potential energy source, fall mainly, but not exclusively, within the so-called seismic regions, which coincide with the margins of active tectonic plate boundaries. Figure 1.2 shows the position of the various plate boundaries. In these areas, the earth's crust is in motion and is being stressed and weakened. Here hot mantle material can rise close enough to the surface so that the transfer of heat to the earth's crust can take place.

The potential areas for geothermal heat are large. Major sources have been located along most of the accessible plate margins. Although the seismic zones are fairly large, the bulk of the land mass of the world is not in the active plate margins. Active areas include the Pacific "rim of fire," which covers the active volcano areas of New

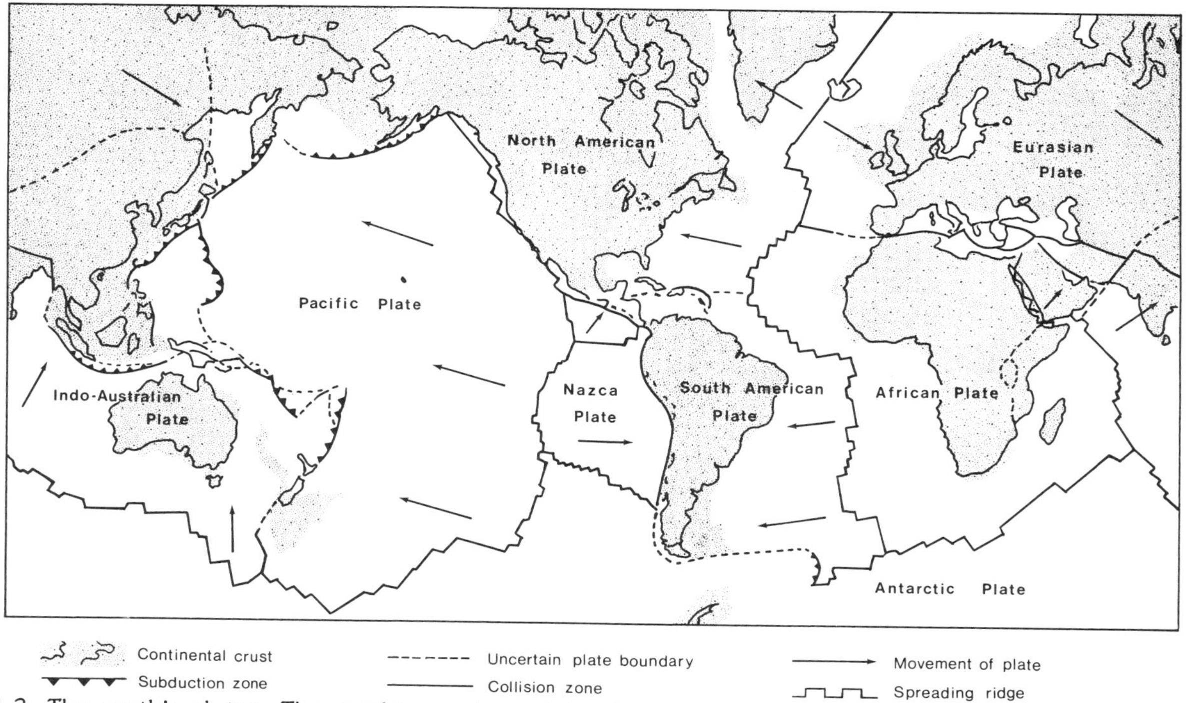

Fig. 1.2. The earth's plates. The earth's crust consists of a set of rigid plates which are moving relative to one another. The resulting stresses at the plate boundaries give rise to earthquakes and active volcanoes. Heat flow and geothermal gradient in these regions is higher than average.

Source: J.D. Garnish, "Progress in Geothermal Energy." Endeavour (New Series Volume 2, No. 2, 1978), p. 67. By permission.

Zealand, Philippines, Papua New Guinea, Japan, Kamchatka (Russia), the west coast of North America, Central America, and the western countries of South America. In the African Rift Valley, steam fields have been identified from Kenya to Ethiopia, while the east-west trending zones include sites in the Azores, Italy, the Balkans, the Middle East, India, Indonesia, and the Caribbean. There are, then, a large number of countries which have volcanoes and other manifestations which indicate geothermal potential.

The potential of geothermal energy, however, extends well beyond the areas where surface manifestations occur. Techniques are now being developed which do not require surface indication of the earth's heat as a starting point for investigations work. Thus, in the future it will be possible to discover geothermal fields where none appeared to exist. This has occurred where drilling for oil has disclosed geothermal fields where surface manifestations were absent. Techniques are being developed which, it has been predicted, will ultimately enable man to obtain earth heat in large quantities at economic costs in all parts of the world.(3) Thus, areas which are presently not included in lists of potential geothermal development areas could in the future be considered to have this exciting energy potential available for development.

Over 80 countries are listed as having geothermal potential. For some of them, such as New Zealand and the Philippines, the resource has become an indispensable part of their total energy supply. Other countries, such as Saudi Arabia with its supply of oil, have little interest in developing their geothermal potential. The countries and regions which have known and theoretically probable geothermal potential are listed in table 1.1.

CLASSIFICATION OF GEOTHERMAL RESOURCES

Geothermal resources may be classified in a number of ways, but basically they fall into three main categories: 1) hydrothermal convection systems, 2) hot igneous systems, and 3) regional conductive environments. Each of these three areas can be broken down again into the following classifications:

1. Hydrothermal convection systems can be identified in three areas:

   a. vapor-dominated systems,
   b. high-temperature liquid-dominated systems (above 150°C; 302°F), and
   c. moderate-temperature liquid-dominated systems (90°C-150°C; 194°F-302°F).

2. Hot igneous systems have two main classifications:

   a. molten part, and
   b. crystallized part (hot dry rock).

Table 1.1. Areas of Known or Probable Geothermal Potential

Area	Classification
Africa (North)	
Algeria	B, C
Morocco	B, C
United Arab Republic	B, C
Africa (Central)	
Cameroon	B
Chad	B
Nigeria	B, C
Virunga volcanoes	A
Africa (East)	
Ethiopia	B, C
Somali Republic	B
Kenya	B
Uganda	B
Rwanda	B
Congo (east)	B
Zambia	B
Mozambique	B, C
Rhodesia	B
Malagasy Republic	B
America (North) (Including Canada, Arctic)	
Christian Island	C
Melville Island	C
Alaska Gulf	A, B
Wrangell Mountains	A, B
Northern Andes	A
California	B, C
Nevada	A
Oregon	A
Washington	A
Wyoming	A, C
Utah	A
Arizona	A
New Mexico	A
Colorado	A
Idaho	A
Dakotas	A
Arkansas	C
Louisiana	C
Oklahoma	C
Texas	C
Gulf of Mexico	C
Rainbow Lake (Canada)	C
Hawaii	A, B
America (Central)	
Mexico	A, B, C
Guatemala	A
El Salvador	A
Honduras	A
Nicaragua	A
Costa Rica	A
Panama	A
British Honduras	A
America (South)	
Columbia	A, C
Venezuela	C
Trinidad	C
Ecuador	A
Peru	A, B
Chile	A, B
Brazil (Andean)	A
Bolivia	A
Paraguay	A
Argentina	A
Galapagos Islands	A
Antarctica	
South Shetlands	A
Graham Land	A
Europe	
Austria	C
France	C
Germany	C
Great Britain (offshore)	C
Holland	C
Hungary	C
Italy	A, B, C
Poland	C
Romania	C
Spain (south coast Canary Islands)	A, B
Far East	
Australia	C
Burma	C
China (eastern provinces)	A, B
China Sea (south)	A, B, C
Bengal (east)	C
India	B, C
Indonesia	A, C
Japan	A, C
New Guinea	A, C
Timor	A, C
Middle East	
Afghanistan	A, B, C
Baluchistan	A, B, C
Pakistan	A, B, C
Persian Gulf	A, B, C
Iran	A, B, C
Israel	B
Jordan	B
Lebanon	B
Saudi Arabia	B, C
Syria	B
Tibetan highlands	B
Turkey	A, B
Island Arcs	
(1) PACIFIC	
Aleutians	A
Fuji-Bonin Zone	A
Halmahera	A
Japan (north and west)	A
Indonesia (Sumatra and Java)	A
Marianas	A
Kamchatka	A
New Britain	A
New Hebrides	A
New Zealand	A
Northern Celebeses	A
Philippines	A
Ryuku Islands	A
Solomon Islands	A
Tonga Kermadec Islands	A
(2) Caribbean	
Lesser Antilles	A
Puerto Rico	A
(3) Eastern Mediterranean	
Aegean Islands	A
Greece	A
Northern Crete	A
Mid-Atlantic Ridge	
Iceland	A
Jan Mayen	A
Spitzbergen	A
Soviet Union	
Aspheron Peninsula	C
Azerbaidzhan	C
Ciscaucasia	C
Turkmeniya	C
Ukraine (east and west)	C
Urals	C
Uzbekistan	C
Volga	C

Classification of Fields:
A: Acid volcanic association
B: High-temperature zones
C: High-pressure reservoirs

Source: C.J. Banwell, "Geothermal Energy and Its Uses: Technical, Economic, Environmental, and Legal Aspects," Proceedings of the Second United Nations Symposium on the Development and Use of Geothermal Resources, San Francisco, Calif. May 20-29, 1975, p. 2265.

3. Regional conductive environments consist of two systems:

   a. geopressured part, and
   b. normal pressured part.(4)

The different types of geothermal resources occur in different regions of the world. While all the resources have development potential, some of the geothermal resource categories are more highly developed than others. Electric power generation in numerous countries uses high-temperature hydrothermal convection systems with temperatures normally in excess of 200°C (392°F). Although occurring in all continents and continental margins, the immense quantity of stored heat of the conduction-dominated environment with its hot water geothermal resource is not very widely used in comparison with its potential. Nonelectrical utilization of this type of hot water is, however, being expanded in various countries, including Iceland, Japan, France, and Hungary. In Russia, experts have stated that about 50 percent of its territory contains geofluid between 50°C and 130°C (122°F and 266°F).(5) In chapter 5, examples of uses of this category of geothermal resource are given.

The heat in the molten parts (magma) of the hot igneous (volcanic) system is not used at present. The associated resource – the heat on the margins of the molten system and in the solidified hot systems – shows promise of wide utilization. Techniques for using the hot dry rock resource are being developed. Methods are being researched to fracture the rock so a large heat exchange can be established and to circulate the heat transfer fluid.

USES OF THE EARTH'S HEAT

Different countries are presently using their geothermal resources in different ways. The possible alternative uses of the earth's heat in development and energy utilization are as wide as the uses made of other heat sources. There have been many broad uses of the earth's heat, but in reality these uses are limited only by man's imagination, ingenuity, and technological skills.

The particular range of methods for using the geothermal resource depends to a large degree on the temperature range of the resource. Figure 1.3 shows examples of the uses for geothermal fluids at various temperatures.

Many different uses are being made of geothermal resources in various fields, including agriculture, industry, business, and tourism. Although each use of the resource will vary according to local needs, a broad classification of actual and currently feasible uses of geothermal energy is given in figure 1.4.

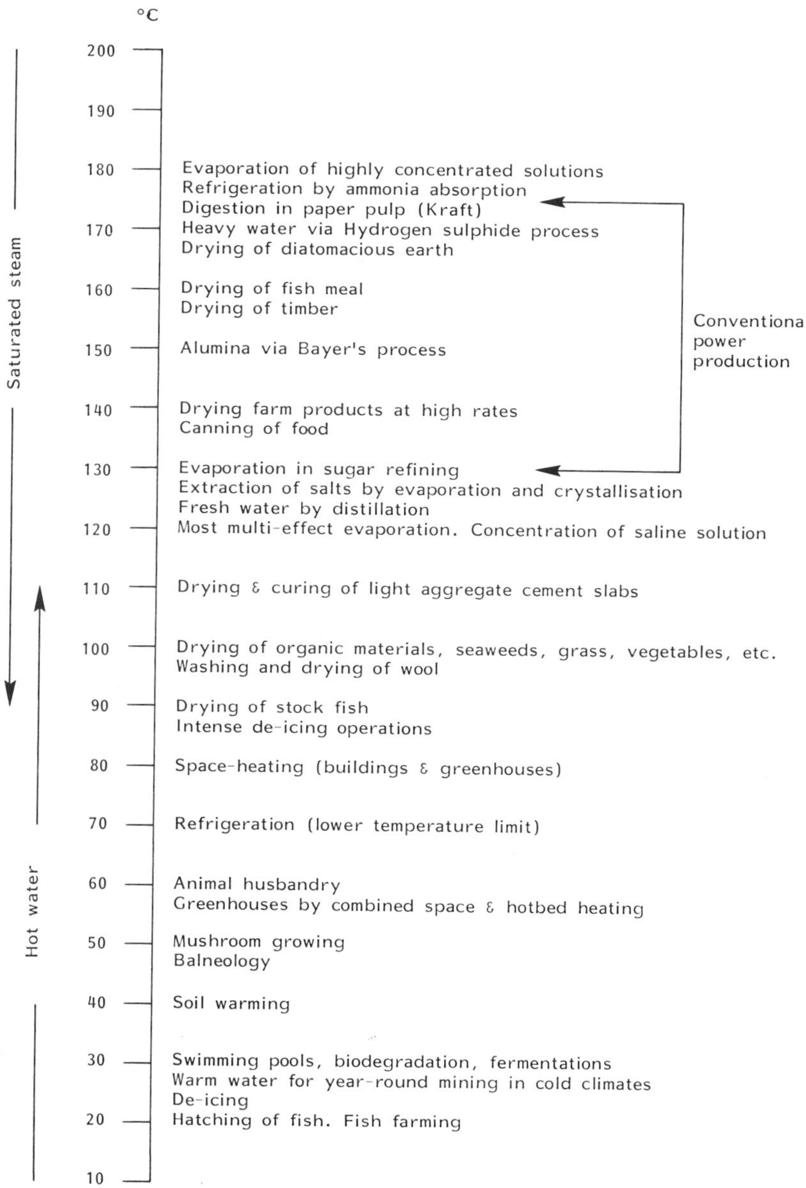

Fig. 1.3. Approximate temperature requirements of geothermal fluids for various applications.
Source: H. Christopher H. Armstead, Geothermal Energy (London: E. & F.N. Spons, 1978), p. 15. Reprinted with permission.

A. Residential and Commercial
1. Space heating
2. Space cooling, air conditioning
3. Domestic hot and cold water
4. Swimming pools
5. Cooking
6. Clothes drying
7. Refrigeration
8. Horticulture
9. Melting snow and ice on driveways and roads; unfreezing of sewer lines in arctic rural areas

B. Balneology/Tourism
1. Health spas
2. Hospitals, sanatoriums
3. Tourist centers and resorts
4. Bottled water

C. Agricultural
1. Greenhouses for vegetables, flowers, plants
2. Hotbed; hothouses
3. Soil warming
4. Animal husbandry – heating of fowl runs, pigsties, cattle breeding
5. Irrigation
6. Water desalination
7. Cooling for food storage
8. Thermal food culture – seaweed, alga
9. Frost protection
10. Plant cooling and humidity control
11. Pest and disease control
12. Fish culture (aquaculture and mariculture) pond culture, cage culture, flowing water culture: ayn, carp, catfish, eels, flatfish, lobster, mussels, oysters, rainbow trout, seabream, shrimp, whitefish, yellow-tail; breeding of alligators and crocodiles

D. Agribusiness
1. Drying – fish meal, grain, hay, paprika, rice, tobacco
2. Canning
3. Refrigeration
4. Freeze drying – coffee, fish, fruit juices, meat, mushrooms, onions, shrimp, soups
5. Brewing and distillation
6. Sugar processing and refining
7. Pasteurization of milk
8. Production of dried milk, casein, and sucrose
9. Seaweed harvesting and manufacture of alginates
10. Cold for the production of protein and vitamin preparations
11. Rice processing
12. Cheese processing
13. Dehydrated potato products

E. Electrical
1. Dry steam
2. Wet steam – low salinity
3. Wet steam – high salinity
4. Low enthalpy – less than 356°F

F. Industrial
1. De-icing
2. Space heating
3. Refrigeration
4. Pulp and paper manufacture
5. Timber seasoning and drying, sawmills
6. Hemp processing
7. Textile processing; wool washing and drying
8. Manufacture of plastic explosives
9. Gasification of coal
10. Hot feed water for solar stills
11. Cold for the production of synthetic rubber
12. Sewage heat treatment

G. Minerals Recovery and Production
1. Diatomite production
2. Chemicals production – ammonia, ammonium carbonate, chloride and sulfate; anhydrous gypsum, borax, boric acid, boron, bromine, calcium chloride, chlorine, dry ice, heavy water, hydrogen for fertilizers and for fuel, hydrogen chloride, iodine, lithium, magnesium, magnesium chloride, methane recovery for peak heating demands, methanol production for fuel, potassium chloride, soda ash, sodium bicarbonate, sodium chloride, sulfur
3. Recovery of trace elements
4. Fermentation – ethyl alcohol, butanol acetone, citric acid
5. Cement drying and curing
6. All-weather mining
7. Minerals recovery – gold, zinc, titanium, silver, lead, tin, beryllium, copper, antimony

Fig. 1.4. Classification scheme for the geothermal industry: actual and currently feasible uses.
Source: Richard E. Peterson and Nabil El-Ramly, "The Worldwide Electric and Nonelectric Geothermal Industry," Geothermal Energy Magazine (November 1975), pp. 5-6.

PRESENT USE OF GEOTHERMAL ENERGY

Advances in drilling technology, particularly as a result of oil exploration, have meant that geothermal resources can be tapped to a level of 3-3½ km (1.86-2.17 miles). Although there were extensive uses of the earth's heat for domestic and health purposes in earlier times, the development of electricity utilizing geothermal sources first took place in 1904. Following this start, progress was slow until the 1950s, when the pace accelerated considerably. The present worldwide concern for new energy sources and the concentration of research on energy development means that progress in the field of geothermal energy has moved ahead.

As the availability of oil becomes more uncertain and its price continues to rise, the virtues of geothermal energy are being discovered by many in the energy-concerned world.

Geothermal energy has a number of major virtues which add to its attractiveness as an alternative energy source for many countries. Six major virtues help account for its present popularity. First, it is an energy source available within the country and therefore offers security in terms of control and lack of outside dependencies. Second, it has, as indicated in figure 1.3, a wide variety of uses, which means it is highly versatile. Third, its resources are immense and, if properly managed, are not in danger of being depleted. Fourth, compared to other forms of energy such as nuclear fission and hydrocarbons, it has few pollution problems. Fifth, although major improvements will be made in the future in utilization techniques and exploration methods, the technology is readily available to develop most forms of the geothermal resource. Sixth, geothermal energy in terms of costs compares favorably with other forms of energy. As oil and other fuel costs increase, the prospect of utilizing the earth's heat becomes more attractive. Table 1.2 illustrates how geothermal energy compares favorably in terms of capital and operating costs with coal, nuclear, and oil energy sources.

Table 1.2. United States Historical Comparative Costs
(in 1976 Dollars)

Cost	Coal	Nuclear	Oil	Geothermal
Capital (installed $/kW)	600	1,000	350	314
Total operating cost (mills/kWh)	34.82	33.20	29.88	22

Source: Ronald C. Barr, "Summary of Section XI Economic and Financial Aspects," *Proceedings of the Second United Nations Symposium on the Development and Use of Geothermal Resources*, San Francisco, Ca., May 20-29, 1975, cxxiii.

While the virtues of geothermal resources brighten the future outlook for the earth's heat, there are scientific and technical problems hindering geothermal energy development. These problems lie mainly in the limited ability to utilize either magma or hot dry rock or the geopressured resource, which constitute the greatest proportion of the geothermal resource. Although research is being conducted in all of these areas, geothermal development depends greatly on very basic advances in science and technology.

The worldwide shortage of energy and the uncertainties about traditional energy supplies mean that nations are treating energy policies and options as a top priority. More resources are therefore going into scientific and technological research in the energy field.

The main use of geothermal resources has been for electrification schemes. Development of electricity-generating capacity from geothermal resources is going on in at least 22 countries; by the mid-1980s, these countries will have a significant electricity output from geothermal sources. Table 1.3 gives details of the installed capacity for the 22 countries as of December 1976, and gives estimated capacity for 1985 and 2000.

As figure 1.3 shows, there are also many nonelectric uses of geothermal energy. From antiquity, man has used the warm waters created by the earth's heat for cooking, bathing, and heating. Direct use of the earth's heat, rather than its conversion to electricity, is in fact the most efficient way to use geothermal resources. Conversion to electricity, however, enables the energy resource to be widely dispersed. It is not possible to detail with accuracy the extent to which geothermal energy is used throughout the world or for nonelectric purposes, but its use in this manner is growing both in terms of capacity used and in methods of application. As the potential, versatility, and relatively low cost of geothermal resources are recognized, the nonelectric use of the resource will increase. For example, geothermal energy is being directly used for a paper and pulp mill, and the heating of government buildings in Rotorua, New Zealand; for a diatomite drying plant in Iceland; for district heating in Iceland, the United States, the Soviet Union, France, and Hungary; and for greenhouse horticulture in Iceland, Hungary, and the Soviet Union. Hungary has a total of 1.9 million square meters (20.4 million square feet) of geothermal greenhouses and plans to have a total of 9 million square meters (96.8 million square feet) by 1990. Additionally, geothermal energy is used to control breeding temperatures in aquaculture facilities growing carp and eel in Japan. At Raft River, Idaho, researchers are experimenting with geothermal aquaculture to raise juvenile channel catfish, tilapia, and giant freshwater shrimp. At Raft River, the geothermal brine is used both to control temperatures and to provide natural chemical nutrients. Some reports indicate that the catfish grow rapidly under geothermally controlled conditions – doubling their weight in 14 days.(6)

Table 1.3. Electricity-Generating Capacity from Geothermal Resources

	1976 Installed Capacity (MW)	1979 Installed Capacity (MW)	1985 Estimated Capacity (MW)	2000 Estimated Capacity (MW)
United States	522	674	3,000-4,000	20,000-40,000
Italy	421	421	800	-
New Zealand	202	202	282	352
Japan	68	218	1,400	48,000
Mexico	78.5	150	400-1,400	1,500-20,000
Soviet Union	5.7	5.7	-	-
Iceland	2.5	34	150	500
Turkey	0.5	0.5	400	1,000
Canada	-	-	10	-
Costa Rica	-	-	100	-
El Salvador	60	60	180	-
Guatemala	-	-	100	-
Honduras	-	-	100	-
Nicaragua	-	-	150-200	300-400
Panama	-	-	60	-
Argentina	-	-	20	-
Portugal	-	-	30	100
Spain	-	-	25	200
Kenya	-	-	30	60-90
Indonesia	-	-	30-100	500-6,000
Philippines	-	223	700	1,000
Taiwan	-	-	50	200
Totals	1,360.20	1,988.20	8.017-10,137	73,712-117,842

THE INTEGRATED PROJECT PLANNING AND MANAGEMENT CYCLE AND GEOTHERMAL DEVELOPMENT PROJECTS

The development of geothermal resources for electrification took its first historical leap forward in Italy in 1904. Prince Piero Ginori Conti first promoted electric power generation using earth heat from the dry steam fields in Larderello. By 1913, a 250-kW power station was in service in Larderello. Today the geothermal resource complex has a capacity of more than 400 MW. Following this pioneering effort, many other geothermal power projects have been developed over the years. This book will examine three important geothermal projects. The first project case history, covered in chapter 2, tells the story of the development of the Wairakei power project in New Zealand. The second geothermal project in the world, this project utilized the wet steam fields situated in Wairakei and pioneered many techniques which were to be used on later geothermal power projects. Completed in 1963, the Wairakei project has a capacity of 192 MW. Chapter 3 analyzes the development of a power project in Hawaii, a pilot project with 3.5 MW that will be operational in 1980 and is expected to lead to further major developments. Chapter 4 considers the Tiwi, Philippines, geothermal project, operating since late 1978, which has a capacity of 55 MW.

Each of the projects is written as a case history cast in the framework of the integrated project planning and management cycle (IPPMC).(7) The IPPMC is a conceptual tool for observing and analyzing the unifying process that constitutes the life of a development project (fig. 1.5). This integrated framework has been developed to clarify the major phases and tasks that constitute the entire spectrum of a given project from planning through implementation, evaluation, and refinement. It can be noted from figure 1.5 that policy issues are inherent in the IPPMC conceptual framework.

The following brief discussion of the IPPMC will provide an understanding of the organizational framework within which each of the geothermal case histories was researched and written. The IPPMC may be divided into four major phases: 1) planning, appraisal, and design; 2) selection, approval, and activation; 3) operation, control, and handover; and 4) evaluation and refinement. Specific tasks may be further identified within these four phases.

Figure 1.5 illustrates the relationships among the phases of the project cycle, the tasks within each of the phases, and the overall dependency on central policy issues. It must be emphasized that the project cycle is an ideal model; not every project will conform exactly to it. The tasks of the cycle, furthermore, are not necessarily sequential – they make take place at the same time or in a different order – nor are all of them necessarily required. For example, the leader of a country might decide that a geothermal scheme is needed for an electric project. He decides to have it built and instructs his subordinates accordingly, thereby bypassing the first two tasks in phase 1 of the cycle. A continual feedback and dependency relationship does exist among the tasks, however. Each task is dependent upon and is influenced by the others.

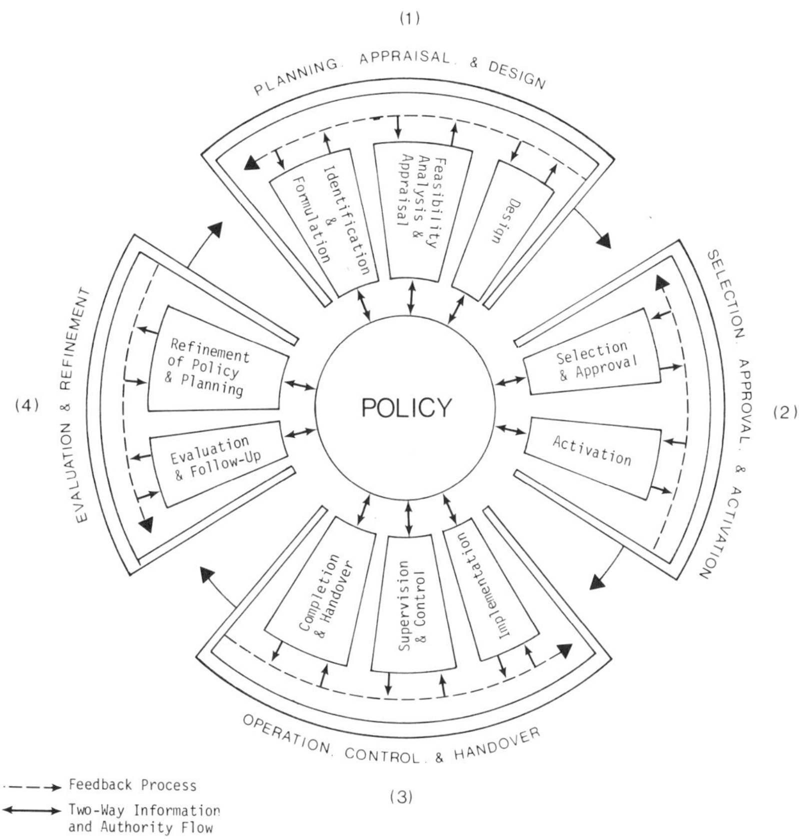

Fig. 1.5. Integrated project planning and management cycle: the four phases.
Source: East West Center, Honolulu, Hawaii.

There is a two-way flow of information between those responsible for policy and those responsible for managing each of the project tasks. This feedback to policy makers and management's response is an important part of the integrated project cycle. Decisions on project implementation, though in the hands of the manager on a day-to-day basis, are closely linked to the policy framework in which the project operates. Thus, all tasks within the four phases of the IPPMC are tied together by policies emanating from the various authorities concerned with the projects.

The IPPMC framework emphasizes the interdependent and cyclical nature of projects. However, because each task within the four phases of the cycle is distinct and must be examined as an individual entity proceeding in an orderly time sequence, the cycle must also reflect this linear progression.

Some projects, it should be noted, may not proceed beyond the first phase. After the tasks within the planning, appraisal, and design phase have been completed, the information fed back to policy makers may lead to a decision to scrap the project. In certain cases, however, for economic, technical, or other reasons, it may not be desirable to stop a project once the first phase has been completed.

With this overview of the integrated project planning and management cycle, each of the four phases and their tasks can be examined in turn.

Phase 1: Planning, Appraisal, and Design

The first phase of the project is planning, appraisal, and design. There are three basic tasks in this phase: 1) identification and formulation, 2) feasibility analysis and appraisal, and 3) design of the project. The first joint task, identification and formulation, involves the actual conception or identification of a project, which may occur in several ways. Basic needs within a country will induce the implementation of projects to satisfy these needs. The planning process often identifies project possibilities for each sector in society.

The major source of projects in developing countries, however, will be the existing departments or ministries, including central planning agencies. Projects may be identified by political parties or government officials. In this case, the motivation to undertake a project may be political, such as an attempt to gain the support of particular constituents. In some countries, private entrepreneurs or multinational corporations will identify projects that meet the criteria established by government.

International agencies have their own procedures for identifying projects. The identification of projects, then, is a process that must take into account various needs, preconditions, and policies if the project idea is to proceed to operational reality.

After a project has been identified, its parameters must be defined. This is part of the formulation task. The formulation of a project involves developing a statement in broad terms which shows the objectives and outputs of the project and also provides an estimate of the various resources required to achieve the project's objectives.

The second set of tasks in the first phase, feasibility analysis and appraisal, are critical ones which in effect involve two distinct operations. A prerequisite of this set of tasks is the development of preliminary designs for the project. The early designs must be detailed enough so that cost estimates and decisions on various aspects of the project can be made.

Feasibility analysis is the process of determining if the project can be implemented. Appraisal is the evaluation of the overall ability of the project to succeed. Projects will proceed to the feasibility stage only if decision makers find them desirable.

While the feasibility analysis and appraisal are being conducted, several critical decisions need to be made. These decisions will determine first, if the project is capable of achieving its objective within the limits imposed by decision makers and second, whether it will proceed. Preliminary estimates of the resources required and basic decisions about size, location, technology, and administrative needs must be made.

Feasibility and appraisal should be approached systematically and deliberately; time spent in researching the feasibility of a project is usually time well spent. Moreover, the findings at this point of the project's life will be useful during other phases of the project, particularly in Phase 3.

Determining project feasibility depends on the accuracy of the information received. Even though the final detailed design of the project can be undertaken only after approval has been given, the preliminary designs form the base upon which future decisions will rest. Most developing countries have to contend with a shortage of both design and research and development capabilities. The result may be a lack of attention to critical aspects of the project. When decisions have been made on the overall project concept, its dimensions and parameters, it is then possible to determine the feasibility of the project in the terms required by the policy makers and the funding agents.

Some projects may require a pilot study as part of the feasibility process. Pilot studies provide data to enable more meaningful decisions to be made about larger projects. The appraisal process may require a comparative study to determine the merits of one project over another. Although the project identified may be feasible to implement, a comparative study determines whether resources are best used in the project or in some other form.

Many governments and international agencies have imposed rigid procedures to be followed when their funds are required. While actual details vary from project to project and from organization to organization, a trend over recent years is for more sophisticated and more systematic project-related studies. For example, to receive a recommendation from the United Nations Development Program (UNDP) for industrial projects, prospective borrowers must undertake market analyses that provide overall national trends in production, foreign trade, consumption, consumer prices, together with details about output type and use, cost of production, and estimated sales. Other agencies have brought in new dimensions to their studies, such as the impact of projects on the social and cultural life of the community, as well as the environmental and ecological impact of the project.

Numerous components of the project can be dealt with in the feasibility report. Studies can relate to the feasibility of the technical, economic, commercial, financial, administrative/managerial, and organizational aspects of the project. Additional political, social, environmental, and cultural factors that affect the project may also be included. Various technical alternatives must also be studied to ensure the suggested approach fulfills project requirements.

Economic studies examine the overall sector into which the project falls and consider how the project fits into the broader sector and the national planning framework. Related to economic feasibility studies, commercial studies may be necessary to determine the overall competitive nature of the proposed project. They will examine the market demands for the output of the project, consider the costs of production, and look at all aspects of the project to determine if it is a viable proposition.

Financial studies determine how much capital is required to complete the project. These studies determine whether the project can sustain its financial obligations, have adequate working capital, and generate enough funds to ensure adequate cash flow to keep the project operational.

Administrative/managerial studies determine the adequacy of procedures to control and direct the project. Studies in this area are not always undertaken, even though all projects would benefit from them. Their objective is to determine whether a project that is economically, financially, and commercially sound can be properly implemented by available managerial and administrative procedures. Many countries suffer from a lack of management and administrative capacity to direct projects. Related to this problem is a lack of ability to ensure that a project can be administered effectively within an appropriate agency or organization. Because administration of a project differs from normal departmental procedures, a careful assessment of the operational methods of existing units is necessary to ensure that a project's unique features can be catered to. Even though a project may be conceived and sponsored by an existing department, the department itself may not be the appropriate body to administer it. This is especially the case when the involvement of a wide group of outside personnel and agencies is necessary, since existing departmental procedures are often unable to provide the necessary flexibility.

Once the feasibility studies have been completed, a meaningful appraisal of the project is possible. Policy and decision makers and lending institutions may carry out the appraisal. They satisfy themselves that the project meets the conditions that enable it to proceed. Their concern is to determine whether or not the project is the best means of reaching the objectives they have set. They may review the project itself and alternative means of reaching the objective.

Potential lending institutions may undertake their appraisal with a healthy skepticism toward all phases of the project. They attempt to determine whether or not the project is intrinsically sound and whether or not all the circumstances that surround it are viable.

The last task within this phase of the integrated project cycle is design. As mentioned earlier, preliminary design criteria must be established before the project feasibility and appraisal task begins. Once it has been determined that the project will continue, the design task proceeds. Design is a critical function. It establishes the basic programs, allocates responsibilities, determines activities and resources, and sets down in operational form the areas of priority and functions to be carried out. All inputs relating to projects, including personnel, skills, technical input, and so on, must be determined at this point. Environmental factors, social criteria, technological requirements, and procedures must be assessed and included.

The design task also includes the preparation of blueprints and specifications for construction, facilities, and equipment. Operating plans and work schedules are prepared and brought together in a formal implementation plan; contingency plans may also be prepared. Designers must bring together the views of policy and decision makers and technical experts in such a way that the design reflects the inputs of all those contributing to the project.

Phase 2: Selection, Approval, and Activation

This phase of the project has two major tasks: 1) selection and approval, and 2) activation. Project selection takes place after the project has been accepted by policy makers and funding organizations as meeting the feasibility criteria. At this point, the design function, including the formal implementation plan, has been completed. The project will be well defined, with key elements identified and the inputs required from organizational personnel, technicians, and outside consultants clearly identified. The selection of one project for implementation over another is made on the basis of several criteria. Policy makers consider the overall feasibility of the project and the priority of the project area. If a project fulfills a major need or contributes to national or sector goals and is politically desirable, it may be selected for implementation over a competing project that is not politically important. Funding agencies, however, have a variety of techniques for determining whether resources will be allocated to a particular project. These techniques may range from cost-benefit analysis to other more complex forms of analysis. The overall requirement, however, is that the policy makers and the funding agency conclude that the project itself has a priority claim for resources required for the project. Therefore, the selection process is normally a competitive one.

The selection of a project for implementation requires negotiations to be undertaken to obtain formal approval from national authorities, funding agencies, and others contributing to the project. This requires the finalization of funding proposals, agreements, contract documents, including tenders and other contracts and the introduction by government or some other organization of appropriate regulations.

Activation of the program involves the coordination and allocation of resources to make the project operational. Activation is a complex process in which the project manager has to bring together an appropriate project team which may include professionals, technicians, and resource personnel. Other contributions to the project may come from other groups, such as outside consultants, contractors, suppliers, and policy makers in other agencies. The outside inputs must be coordinated with the work of the project team. Responsibility and authority for executing the project must be assigned at this point. This will include the granting of authority to make decisions in areas relating to personnel, legal, financial, organizational, procurement, and administration matters.

The activation task must ensure that planning for all phases is undertaken so that delays in vital inputs do not occur. Organizational and administrative procedures, together with feedback and response to policy makers' decisions, will have an important bearing on implementation. Concern for detail and proper planning during activation can save a great deal of time and resources during later phases of the project. At this point, the actual work of the project is about to begin.

Phase 3: Operation, Control, and Handover

Looking at the development project from the outside, the uninitiated observer might mistake this most visible phase for the entire project itself. As has been indicated, Phase 3 in fact makes up only a small part of the integrated project cycle. This phase of the project has three sets of tasks: 1) implementation, 2) supervision and control, and 3) completion and handover.

Implementation involves the allocation of tasks to groups within the project organization. Implementation of the project will be based on procedures set down during the two earlier phases. At this point, a final review of the project design and timetable will be undertaken, and any necessary changes or adjustments will be included. Decisions about the procurement of equipment, resources, and manpower also need to be made. Schedules and time frames need to be established; efficient feedback, communication, and other management information systems must be set up. The responsibility for implementation falls within the jurisdiction of the project manager. The project manager will need to work with policy makers, authorities, and organizations related to the project as well as with policy makers controlling the project. His task is a complex one, requiring him to steer the project through many obstacles.

The second set of tasks in Phase 3 is supervision and control. Supervision and control procedures must be activated to provide feedback to both the policy makers and the project manager. Control procedures must identify and isolate problem areas; the limited time span of a project means that fast action is necessary if costly delays are to be avoided. At this point, specific management tools, such as the critical path method (CPM), the program review and evaluation techniques (PERT), and other forms of network analysis, are particularly useful. These control and supervision techniques break down a project into detailed activities and establish the interrelationships among the various activities. This allows the project manager to organize the project into manageable components, to coordinate all activities, and to set a time-sequence schedule for project implementation. Although using such techniques means spending more time prior to implementation, it is time well spent. Not only will these techniques give the project internal coherence, but they will also save implementation time by isolating any problems to the appropriate project component.

In addition to internal control, those providing funding for projects will maintain an independent monitoring and control system for the project. The project manager will therefore have to meet control criteria established by either the government or another controlling agency, or perhaps by the funding institution. This may involve using specified procedures, such as international competitive bidding, for supply contracts. Formal procedures are established by many international organizations for the procurement and control of resources.

Whatever supervision and control techniques are used, they must take into account the changing patterns that occur during the life of the project. These may include changes within the policy and political structures, difficulties with procurement, and poor performances by project team members and contractors. In many cases, the overall project design will need to be reviewed. Many technicians are involved in the supervision and control processes, and adequate information flow in all directions – from the project manager and from those within his organization assigned special responsibilities – is essential if these procedures are to be effective. As part of supervision and control, any problems relating to environmental factors must also be identified and appropriate action taken.

Control procedures are useful only if action is taken to correct any deviation. It should also be noted that both personnel and input patterns change naturally as the project proceeds through its four phases. As work on some tasks is completed, other personnel, experts, and contractors move in to begin new tasks. Personnel must adjust to their new environment, and procedures need to be reviewed and updated to meet the changing situation.

Project completion prepares the project for phasing out and handover to another form of administration. These constitute the third task of this phase. Project completion consists of scaling down and dismantling the project organization. It also involves the transfer of project personnel to other areas of operation. Assets and other facilities, including equipment and technology, may not be required by the operational project. Provision for their transference must be made, since it is not always possible to have an automatic transition from the development to the operational stage.

The process of completion may take place over a considerable period. As various parts of a project are completed, however, they may be taken over by a new organization, and handover may therefore be accomplished in a piecemeal manner. It is essential that development resource linkages between scaled-down projects and those projects in the elementary stages of implementation be planned systematically to ensure optimal use of limited project resources, particularly in the context of broader development programs. The new project, when operational, will have an effect on other aspects within the sector. As the project becomes operational, the new controlling organization must have the skills, personnel, and technical backup required. Key personnel working in the development stage will often transfer over to the new controlling organization.

In cases where technical, financial, political, or other factors prevent projects from being completed according to the original terms, handover and termination procedures may have to be implemented at an earlier stage. This may involve considerable loss as far as the project is concerned. In this situation, the objective should be to liquidate the project in a way that will obtain the most benefit.

As a project nears completion, special reporting systems should be set up so that full information relating to the project is available. Completion reports will be prepared for various authorities, including funding organizations and policy makers.

The actual handover of the operation of the project involves finalization of contracts, termination of loan facilities, and so on. Handover also includes the transfer of the project activity and resources to the new administration. This is a critical task. While the development of the project can be viewed initially as a creative phase, once the project is completed, it must be viewed as a long-term operational program.

Phase 4: Evaluation and Refinement

The final phase of the project is the evaluation and refinement of policy and planning factors. The first task is evaluation and follow-up. While it is possible to evaluate project results immediately, actual benefits – both anticipated and unanticipated, together with side effects – may not become apparent until the project has been operating for some time. Evaluation thus needs to cover several time periods. Evaluation normally includes a retrospective examination of the project in attaining its intended goals within the framework of both the timetable and the budget. However, experience clearly demonstrates that it is necessary to consider evaluation as an ongoing process integrated with each phase of the IPPMC. For example, evaluation procedures must be designed to analyze and propose solutions to problems that may arise during the tasks of activation, implementation, supervision, and control. Ongoing evaluation, which includes retrospective evaluation, should result in a careful documentation of experiences that can provide both insights and lessons for improving project planning and project management in the future.

Evaluation of a project can take several forms. These include evaluation by those responsible for implementing the project and by others with an interest in the project, including funding organizations and contractors. Those funding the project will undertake a thorough investigation of its financial aspects, including an effectiveness study of goal attainment. The agency responsible for the project will be concerned with determining whether goals have been attained and whether the expected impact on a sector or on national development will be achieved. The studies should also consider, in addition to impact on the target group, the impact of the project on the political, social, cultural, and environmental factors relating to the project. An exhaustive evaluation of each phase to determine its contribution to the project in regard to budget, timetable, and other factors is most desirable. In most cases, however, the project as a whole is evaluated with little effort made to analyze each phase or each task separately.

International agencies, such as the World Bank and the United Nations, have their own procedures for evaluating projects. These may be useful to policy makers, since they provide the opportunity for comparative analysis with similar projects.

Related to and often arising from the evaluation of a project is the need for project follow-up. Follow-up activities may vary from determining how unmet needs can be satisfied to action on project tasks not properly fulfilled. The piggyback or follow-up projects mentioned earlier may come into play at this point. For a project to achieve its full objective, smaller or related projects may need to be implemented almost immediately. There is then a clear need to relate follow-up action closely to evaluation of projects. Follow-up action is one aspect of the project manager's role which could involve considerably more commitment than he initially envisaged. If follow-up action means the difference between the project's being fully operational or not, then it is a wise investment to undertake these activities as quickly as possible. Aspects arising from the follow-up procedures may be useful in the future. If the project is successful, guidelines can be set down for the project to be repeated in another setting.

The second and last task is refinement of policy and planning. Policy makers and managers will need to refine their procedures in the light of each completed project. Experiences and lessons learned should be the foundation on which planning and policy tasks are reviewed. As the essential controlling force, policy procedures must be continually updated to meet challenges in the future. Planning must also be able to meet any new demands or new situations. Refinement of these procedures is an important contribution that the project can make to future development programs.

The IPPMC is a flexible model for all phases of a project from conception through completion. The cohesive force unifying all the phases and tasks of the IPPMC is the power and authority relationship vested in various policy makers, ranging from top government and political decision makers to those in charge of one aspect of the project. The project manager, the staff, and those contributing to the project as consultants or contractors are bound by and exist within the framework of policy decisions. Analysis of these changing relationships through the IPPMC model can provide a comprehensive overview of a development project. It is also useful for policy makers in providing guidelines for addressing policy issues as a basis for more viable policy formulation and related decision making. The three geothermal case histories in this book highlight the need for a sound interface between policy makers and those responsible for managing the implementation of the geothermal projects. The cases also highlight how interaction and interrelationship between the phases and tasks occurs within the IPPMC framework. Each of the geothermal projects documented operates from within a different political and social environment. Each project, however, follows a similar path in its quest to develop the geothermal resource.

NOTES

(1) See H. Christopher H. Armstead, Geothermal Energy (London: E. and F.N. Spon, 1978), pp. 31-37, for estimates of earth's heat potential.

(2) Joseph Barnea, "The Earth and Its Unknown Interior," Geothermal Energy 5 (February 1977): 22.

(3) H. Christopher H. Armstead, "Geothermal Energy Development," Progress in Energy Combustion Science, vol. 12 (Oxford: Pergamon Press, 1977), pp. 181-231.

(4) See D.E. White and D.O. Williams, "Assessment of the Geothermal Resources of the United States – 1975, United States Geological Survey, Circular no. 276 (1975): 147.

(5) M. Tikhonov and I.M. Dvorov, "Proceedings of the United States Nations Symposium on Development Utilization of Geothermal Resources, Pisa," Geothermics, special issue 2 (1970): 1102.

(6) L.J. Goodman, T. Miyabara, and B. Yount, "Broad Issues of Geothermal Development" (Paper presented at Department of Energy Workshop on Geothermal Overview for the state of Hawaii, May 9-12, 1979).

(7) For a complete discussion of the IPPMC, see Louis J. Goodman and Ralph Ngatata Love, eds., Management of Development Projects: An

International Case Study Approach (New York: Pergamon Press, 1979), and Project Planning and Management: An Integrated Approach (New York: Pergamon Press, 1980).

2 Wairakei Geothermal Power Project: New Zealand

Ralph N. Love
Richard Bolton

PROJECT BACKGROUND

A belt of geothermal energy in the North Island of New Zealand stretches from White Island off the coast of the Bay of Plenty to Mount Ruapehu, covering an area about 150 miles (240 kilometers) long and 30 miles (48 kilometers) wide (fig. 2.1).

Maori mythology explains the origin of the thermal activity as the creation of the god Ngatoro-i-rangi. Charmed by its beauty, Ngatoro decided to climb snow-capped Mount Tongariro. Before he began the first ascent of the mountain, he commanded his followers not to eat during the time he was gone as their abstinence would give him strength. But his followers became impatient after a long wait and lit their cooking fires. As they ate, Ngatoro became colder and colder until he could climb no further. Near the summit, almost frozen to death, he prayed to the gods on Hawaiki, the home of the Maori before their migration to New Zealand. The gods heard Ngatoro's prayer and dispatched the fire demons Te Pupu and Te Hoata who plunged into the sea and swam until they reached Whakaari, or White Island. There they raised their heads into the air to get their bearings and the earth burst into flames which have burned ever since. The demons surfaced again for a moment at Motuhara (Whale Island), Rotorua, Tarawera, Taupo, and other sites, including Wairakei. Their flames roared through the holes at each place; they shot through the interior of Tongariro and came out at the summit as Ngatoro lay near death. The intense volcanic heat they brought revived him and he descended safely.

At the time of the first settlement by Europeans, the Maori people living in the area were making extensive use of the thermal activity. The steam was used for cooking and warmth and the hot water for washing, bathing, and healing. For many years the natural geothermal activity remained only a tourist attraction, although various suggestions were made about developing it to supply energy. Following World War

Fig. 2.1. North Island of New Zealand.

II, severe electricity shortages rekindled interest in geothermal energy and serious investigations into the possibility of developing it were begun.

PHASE 1: PLANNING, APPRAISAL, AND DESIGN

Identification and Formulation

The Wairakei Geothermal Power Project is located on the Waikato River in the center of the North Island of New Zealand, about 5 miles (8 kilometers) from the river's outlet at Lake Taupo. Wairakei was chosen for the geothermal development site because it contained many surface manifestations of thermal activity, ranging from pools of hot water to fumaroles, geysers, and boiling mud pools. Superheated steam was being emitted from the ground at the Karapiti fumarole. Moreover, previously the area had been drilled to about 300 feet (91 meters) in an attempt to obtain steam to heat a nearby tourist hotel. The early attempt had failed, however, primarily because the drilling techniques were inadequate for the conditions.

Further reasons for choosing Wairakei as the project site over other sites such as Rotorua included the following:

1. The area contained streams which could provide water needed for drilling.
2. The nearby Waikato River could provide an ample source of cooling water for the project when it reached the production stage.
3. The land belonged to the government.
4. The areas were made up of unproductive wasteland.
5. The location was fairly isolated and development would not detract from the scenic and thermal attractions which were major tourist resources.
6. No private interests were involved.

Preliminary work

Geological surveys and chemical investigations of the thermal waters were carried out in the Wairakei area in the 1930s. The first report on geological factors was prepared in 1937. It described a thermal belt coinciding with a graben extending down thousands of feet, which was filled with sedimentary deposits, probably of marine and lacustrin origin, overlaid with light deposits of volcanic and fluviatile origin. Faulting of the flanks, or at the base of the graben, provided then, and continues to provide today, a path by which magmatic steam infiltrates to the upper formations, where migration of the steam from the fractures through previous beds is thought to be the means of distribution of the steam in the Wairakei area (fig. 2.2).

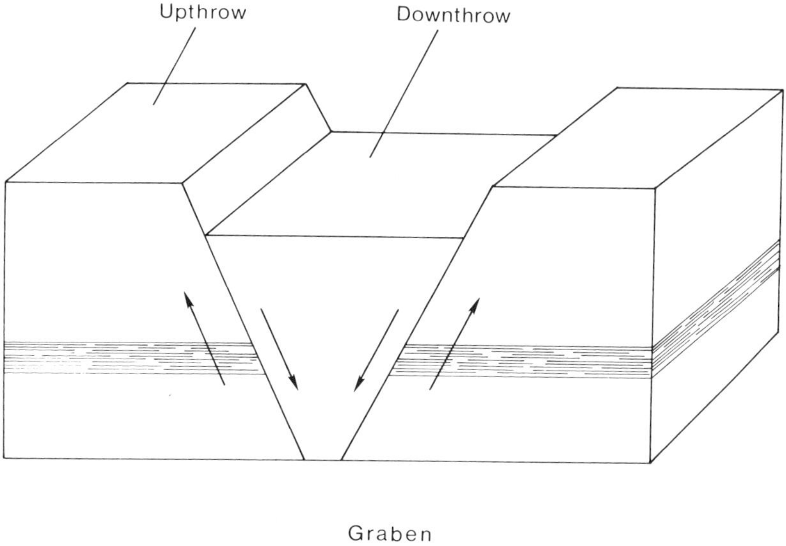

Fig. 2.2. Graben. Formation illustrating the depressed segment of the earth's crust bounded by faults.

Investigations of geothermal steam were considered by the New Zealand Scientific and Industrial Council early in 1940. Recommendations to investigate the problem of utilizing natural heat were made, but at that time it was considered that the North Island was well provided with water resources for power generation and there was little need to consider the possibility of power generation for some time to come. A decision to develop geothermal resources for power generation was therefore put off.

The natural resources were, as in pre-European days, already being used by the local inhabitants. In Rotorua, a major tourist city, heated ground water was being tapped at relatively shallow depths from a number of wells which had been sunk to provide central heating and hot water. The local city council building, a movie theater, a primary school, several hotels, and a number of private dwellings were utilizing this natural heat by the 1940s. In 1944, over 50 wells had been sunk in the Rotorua area and this was increasing at a steady rate. For many years, the hot waters had been used as a health treatment and people came from many parts of New Zealand and from around the world to use them in the belief that the hot waters had special powers that would heal them of various maladies.

Although there was no immediate intention to develop the natural resource, a geologist was appointed by the Department of Scientific and Industrial Research (DSIR) in 1945 to act as a volcanologist for the region. He was to commence a systematic survey of the thermal zone. A conference of DSIR officers was held at Rotorua in January 1946 to discuss the program of investigation to be undertaken. One suggestion was made at the conference which was to have a major impact on the future development of the thermal and geothermal resources of New Zealand. A scientist speculated that the natural steam might be used for the production of heavy water by fractional distillation methods. There was a worldwide shortage of heavy water, an essential ingredient for the budding atomic energy industry. This suggestion was later to attract the interest of the government of the United Kingdom.

Energy for rural development

During the 1930s and early 1940s, the geothermal resources of New Zealand were of scientific interest but of no real practical benefit in terms of future power generation. The situation in New Zealand regarding the demand for power changed dramatically following World War II. By 1947, the fact that New Zealand was having problems supplying the country with needed energy was acknowledged by the government.

New Zealand developed as an agricultural nation and has always been dependent on its rural communities. Its economy is based entirely on agricultural activities, with over 90 percent of its exports based on the processing of agricultural products. The main market for New Zealand farm produce at the time the project commenced was Great Britain. Changes in consumer demand in Great Britain following World War II required New Zealand to increase its output and to place greater emphasis on further processing of its agricultural products. To increase output, farmers had to improve their productivity by electrification and mechanization. It was essential that the rural community have electricity so that farm work such as milking, irrigation, and shearing could be mechanized and output increased. Efficient processing of such agricultural products as cheese and butter was hampered by the lack of power in rural communities.

In his report to Parliament in 1947, the Honorable Mr. R. Semple, minister in charge of the State Hydro-Electric Department, said that in line with the government's policy of encouraging local manufacture wherever possible, the demand for electricity had continued to grow while lack of rain for the past two summers had seriously affected the output of hydro projects. Because of worldwide disorganization in the aftermath of the war, there were serious delays in obtaining materials and skilled staff to build more generating plants to keep pace with increased demand. The problem was not unique to New Zealand. Shortages of electricity and delays in providing additional generating plants were common in England, Australia, Canada, Switzerland, and practically everywhere else electricity was used.

Other energy problems faced the government. Before Parliament, the minister pointed out that New Zealand's power needs were so great that any generating plant added to the system had to be capable of continuous operation over a long period. This, he said, precluded the consideration of diesel-electric plants, except for installation in connection with some particular factory. Other energy options were also rejected. The minister made the point that the cost of fuel-driven plants was very high and that coal was in short supply, so that there was no possibility of obtaining large quantities of coal to generate electric power. Clearly, the growing country was facing an energy crisis. Concluding his report, the minister looked to another alternative – natural steam. He proposed to investigate this matter without delay and to put down trial bores in suitable locations. He said that natural steam would provide power comparable to water power in that it would be supplied by nature. "Under present world conditions," he concluded, "it is essential that the electric supply of this country be independent of imported fuel, and to this end, we should develop the natural resources with which the country has been generously endowed."(1)

In the past, water power through the hydro projects had proved to be extremely efficient and capable of meeting the country's energy needs. The only other major indigenous resource available at that time was geothermal energy. It was, therefore, the 1947 energy crisis that prompted the government of New Zealand to look at alternative sources of indigenous energy. The result was a decision to go ahead with the Wairakei project. Although geological studies had been made and a volcanologist appointed, only scattered information was available on the geothermal belt at the time the decision was made to go ahead with an investigation of its potential development.

The first problem facing those responsible for implementing such an investigation was to bring together, into a network of cooperating agencies, the various government departments and institutions necessary to develop the natural resource.

Government policies and framework

The responsibility for generating and distributing electricity in New Zealand at the time the development of Wairakei was contemplated rested mainly with the government, and this still holds true today. New Zealand has no private utilities; rather, a small number of local bodies are permitted to generate some electricity, while one or two major industries can, in special circumstances, generate electricity for their own use. From the early stages of the project, three government departments were concerned with the development of geothermal power at Wairakei. The Department of Scientific and Industrial Research (DSIR) was one of the earliest. During the 1920s and 1930s, scientific investigations were carried out more as a means of understanding the overall phenomenon than as a serious attempt to determine if natural steam was a viable source of power. One notable exception was a report from the DSIR, written in the early 1930s, which suggested that an attempt should be made to outline "the future of the development and utilization of subterranean heat, emphasizing the great possibilities of this source of natural energy." The DSIR quoted examples of thermal heat projects which had been developed in Italy at Larderello and in Iceland.

The second major department concerned with the development of the Wairakei project was the Ministry of Works and Development (MWD). The Ministry of Works and Development was responsible for a wide variety of development activities in the country and worked closely with the DSIR. The Ministry of Works and Development was, in effect, the government's construction agency and implemented the major development projects in hydro, roads, and other areas. It played the role of both a contractor and of a planning and development agency.

The third department involved at Wairakei was the State Hydro-Electric Department (SHED). This department had the prime responsibility of supplying power to New Zealand. It was concerned not only with the planning but also with the actual running of the power projects operating in the country. SHED had only limited resources to offer the development and construction processes, but it did have major facilities and years of experience in running projects, once they had been established. The Mines Department was also involved in a peripheral role: it had drilling expertise and equipment which was vital to the project.

These three departments, MWD, DSIR, and SHED, had to integrate their skills to develop what was for all of them a new process involving a large number of new technologies; no one department had the capability to undertake the project by itself. These government departments involved in the geothermal project had a history of working together, but the project was one which required a particularly close integration of roles. It was, in effect, a new type of working relationship for most of the departments. One of the difficulties encountered during the early phases of the project was determining who had responsibility for certain key areas of study and development.

The approach taken to coordinate the various government departments was to establish a Geothermal Advisory Committee (GAC), which included representatives from the Ministry of Works and Development, the Department of Scientific and Industrial Research, and the State Hydro-Electric Department. The GAC was established formally by authority of the cabinet on November 23, 1949. Prior to this, two unofficial meetings were held. It is important to note that the GAC was set up with the cabinet's approval and was not an ad hoc official's committee. In effect, this meant the GAC reported to the cabinet through the minister of the DSIR, and not each department through its own ministry.

The first meeting of the GAC was held in September, 1949. It centered on a study of the work which had been undertaken by the DSIR in earlier years and also considered what steps had to be taken to get a major study under way for prospecting and developing the natural steam in New Zealand.

The second meeting of the GAC was held two months later. Again, committee members considered the means of approaching the exploration for geothermal steam in the Wairakei region. At the meeting, geological and geophysical reports on the characteristics of the region were given.

The conclusion of the GAC was that the region was one which merited a search for steam power with reasonable possibilities of success. A theme which continually ran through the discussions was that the government was eager to obtain results and wanted to determine quickly if it was economical to install steam plants to generate electric power. One task of the GAC was to obtain a speedy answer to this government request. Consequently, the early part of the work of the development project centered on the urgent need to determine the economics and feasibility of geothermal steam as compared with water power and other power resources.

Energy alternatives

As a result of the 1947 crisis, the minister of the State Hydro-Electric Department had considered the use of natural steam as a source of power. In 1948, the problem of energy scarcity had not improved. Hydro projects were still the major supplier of power, but poor rainfall had again created problems. The shortage of water was aggravated by a shortage of both coal and oil for steam stations in Auckland and Wellington. In 1948, the shortage of oil fuel was worldwide. As far as the New Zealand government was concerned, there was no reliable information as to the possible duration, or ultimate extent, of the shortage. The minister said, "The fact that sufficient oil is not available for even the relatively small amount of steam generating plants found in the country is very disturbing, as it practically rules out any large-scale use of this fuel in New Zealand for future power generation." Commenting on coal supplies, the minister said, "For one reason or another, coal supplies have not been equal to requirements. The fact is that a regular quantity of coal is not available, and this must arouse grave doubts as to the advisability of the use of coal-burning plants where a better alternative is more readily available."(2) The government was therefore faced with finding alternative energy sources. In his report the minister stated, on rather a prophetic and optimistic note, that the DSIR was undertaking investigations on the use of geothermal steam. He said, "This alternative is attractive provided the conditions are such as to render the development of electric power from natural steam an economic and technical possibility."(3)

New Zealand was then faced with a number of alternatives. There were four possible methods being considered to provide power on a sufficiently large scale to meet New Zealand's needs. These were:

1. Steam power stations using coal or oil fuel,
2. Steam power stations using geothermal steam,
3. Steam power stations using atomic fuel (very uncertain), and
4. Hydro power stations in the South Island with cables across Cook Strait to the North Island

Of the four, the second and fourth alternatives were felt, in 1949, to be technically possible and probably more economically feasible than using coal, oil, or atomic fuel to run power stations.

Investigation and Organization

In November, 1949, the Geothermal Advisory Committee felt that it was essential to continue surveys over wide areas. Because of the pressures to develop quickly an understanding of the potential of geothermal steam for power generation, however, a short-term approach was needed, with exploration concentrated on the Wairakei region. Early estimates had indicated that a five-year program of study would be best. But the GAC felt that intensive survey work could probably provide an answer in 12 to 18 months about the possibilities of using steam to generate power.

To enable the GAC to undertake this work, recommendations were made to the government. These were intended to expedite the geophysical survey. The cabinet was asked to approve the following:

1. That a survey camp be established in the Wairakei area and $21,000 be provided for camp equipment;
2. That one shooting truck, one water truck, and one dynamite storage trailer be provided to enable seismic and drilling investigations to be undertaken; and
3. That a thermal resource exploration committee, comprising members of the Ministry of Works and Development, Hydro-Electric Department, and Department of Scientific and Industrial Research, be appointed to coordinate activities in the Wairakei area.

Committee members hoped that within a period of 12 to 18 months enough reliable information would be available to indicate where boring could take place in the search for steam in quantities large enough to generate sufficient power.

The recommendation of the GAC to undertake the studies on the potential for geothermal steam was approved by the cabinet in late November, 1949. The GAC felt that because there was pressure to complete the investigation at Wairakei within a short time, a decision had to be made on the purchase of drilling equipment. It would be necessary to put this equipment into working order and for a drill crew to acquire some experience in operating it. This would take time. An oil company had a large drilling rig for sale. The GAC felt that its price of $140,000 was reasonable, considering that to purchase a drill from the United States would cost perhaps double that sum.

It was evident that the bulk of the actual work would be done by the Ministry of Works and Development, while the Department of Scientific and Industrial Research would provide major input on the scientific side of the geothermal development. The Commissioner of Works, the permanent head of the MWD, would have responsibility for carrying out the work. Accordingly, the Commissioner of Works asked the GAC to consider a proposal to put the work under the MWD staff already working in the area on hydro investigations.

At a December, 1949 meeting convened to discuss this proposal, the organizational framework was set forth by the general manager of the

SHED. It was not clear why he should be the one to do this, as representatives of the MWD were present. The record of the meeting prepared by the secretary of the DSIR showed that the general manager of SHED's explanation of the organizational framework was completely misunderstood. The record shows that the GAC believed the hydro investigations on the Waikato River were controlled by an investigating committee comprising members of the MWD and SHED, which was not true. Accordingly, they resolved that the geothermal investigations at Wairakei should be controlled by an investigating committee comprising members of the DSIR and MWD. This was, of course, contrary to the intention of the Commissioner of Works.

The assistant engineer-in-chief was advised in late January 1950 that he and the MWD inspecting engineer were to take charge of the work of investigating geothermal energy for power development. After examining the record of the December 1949 meeting, he immediately realized the misconception that had arisen, and advised the engineer-in-chief of the MWD accordingly and at some length.

The comments made were discussed by the GAC and by the heads of the three departments, the MWD, DSIR, and SHED. The Commissioner of Works then wrote to the secretary of the DSIR pointing out the misconception in the notes on the meeting of December 15, 1949 and said that the proposition he had intended to be discussed at that meeting was that drilling and associated work be under the MWD investigations organization at Mangakino.

The engineer-in-chief subsequently wrote to the DSIR confirming the various discussions that had been held and acknowledging that the work could go forward under his general administration. The work was defined broadly in two categories:

1. The investigation for geothermal steam on an orthodox scientific basis, and
2. The drilling of an area in the Wairakei region in the hope of striking steam.

The DSIR was responsible for the first task. The geological survey was to investigate an area about 12 miles (31 kilometers) square encompassing the Taupo-Wairakei region. Their relation with the MWD would be exactly the same as it was during investigations for other power projects in the hydroelectric field which involved consultations and progress reports from time to time.

As for the second task, drilling was to commence immediately. Drilling would be under the control of the project engineer of the MWD, Mangakino. Initially, the drilling was to be confined to shallow holes, for which the project engineer was to seek the advice of the geological survey. If successful, larger wells would be drilled, for which purpose a rig located in the Hutt Valley was being examined.

The secretary of the DSIR stated that his department was in complete agreement with the general administration proposed for the geothermal steam investigations at Wairakei.

One of the first decisions which had to be made by the Geothermal Advisory Committee was how deep the drilling would have to go. Geological and geophysical investigations had not yet determined to what depth it would be necessary to bore. The GAC was aware that at Larderello, Italy bores had been put down to depths of 2,000 feet (609 meters) and later to 6,000 feet (1,829 meters). The Ministry of Works and Development had a drill which would go down to 1,500 feet (457 meters), two Sullivan drills at the DSIR would go down to 800 feet (243 meters), and the Mines Department had one Franck's drill which could go down to 3,000 feet (914 meters).

The GAC decided to get the three drills on the job as soon as possible. Two Sullivan drills would be used on seismic exploration and the Ministry of Works and Development's drill would be used to drill a bore to 1,500 feet (457 meters). They hoped that a bore drilled down to 1,500 feet (457 meters) would determine the geological structure of rock types to that depth and reveal the possibility of developing steam.

Field organization and administration

The government's urgent desire for early results required a speedy implementation of the decision of the GAC regarding control of the work in the field. Accordingly, on March 9, 1950, a meeting was held at the Wairakei Hotel to discuss the administrative procedures and future program for the investigation of the Wairakei-Taupo area for geothermal steam. The chairman of the meeting was the engineer-in-chief of the ministry of Works and Development. Also present were engineers from the MWD together with scientists, geophysicists, and chemists from the DSIR, and a drill superintendent from the Mines Department. The engineer-in-chief of the Ministry of Works and Development explained that the administrative procedures for the study would be those adopted in the past with hydroelectric investigation work undertaken by the Ministry of Works and Development. The contact between the DSIR and the MWD in the field would be maintained by the superintending geologist of the DSIR in Rotorua, and the project engineer of the Ministry of Works, Mangakino. Drilling equipment and survey staff under the control of the investigation engineer at Mangakino would be utilized for the work as much as possible. A camp was to be built in Taupo to accommodate the MWD and DSIR staff and workmen.

The chairman made clear to those present at the meeting that the minister of Works and Development expected investigation work to be carried out as rapidly as possible, the goal being to obtain concrete evidence in 12 to 18 months about whether or not geothermal steam suitable for electrical power generation could become a reality.

The Mines Department had made their drilling superintendent available to supervise all drill prospecting for steam and he would control the drilling. It was agreed that a report would be submitted at convenient intervals on the geophysical and geological work, along with the surveying and drilling work, by the various officers in charge of each segment of the project. As suitable drilling equipment became available from investigation work being carried out by the Ministry of Works and Development at other locations, it would be made available for geothermal investigation work, augmenting the two drills already being used for geophysical surveys.

The meeting resulted in a general plan of operation that was to be accomplished in three phases:

1. The geophysical survey work would be done with two drilling rigs operated by the DSIR with assistance by MWD survey parties.
2. A row of prospecting holes was to be drilled immediately to test steam resources within the capacity of the drilling rigs already being used by the Ministry of Works and Development. These rigs would be made adequate for steam drilling purposes and be operated under the supervision of the Mines Department drill superintendent. The holes would be spaced at quarter-mile intervals extending from the river below the Wairakei Hotel up to the Atiamuri Road and across to the Wairakei Valley above the geysers. This would give a line of holes right across the thermal area. The holes would be cored for much of their length, with temperature and thermal conductivity measured from core samples. Chemical analyses would also be carried out at this stage (see fig. 2.3 for locations and areas).
3. After correlating the data already available and the data to be made available in the near future from all sections concerned, a site would be chosen for the large Mines Department drill which would put down a deep prospecting bore under the direction of the Mines Department drill superintendent.

The plan of operation had been set up and the administrative procedures laid out. The GAC planned that if the three phases of the investigation were followed, a report could then be made within 12 to 18 months for the government on the real possibilities of generating electrical power from geothermal steam.

Investigation reports

Investigation work commenced in February 1950 with the guidelines for the areas of investigation clearly drawn. The DSIR research covered the chemical and physical investigations, plus the geological, seismic, magnetic, gravimetric, and radioactive surveys, and the petrological examination of drill hole cores. The Ministry of Works and Development was directly responsible for the drilling program and utilized experienced personnel from the Mines Department. The administration and coordination of the overall work was the responsibility of the Ministry of Works and Development.

Progress reports by the end of 1950 had been made on every phase of the investigation. But after ten months, there were still no definite conclusions about the actual power potential of the field. The MWD and the Minister of Works and Development hoped and expected that by the

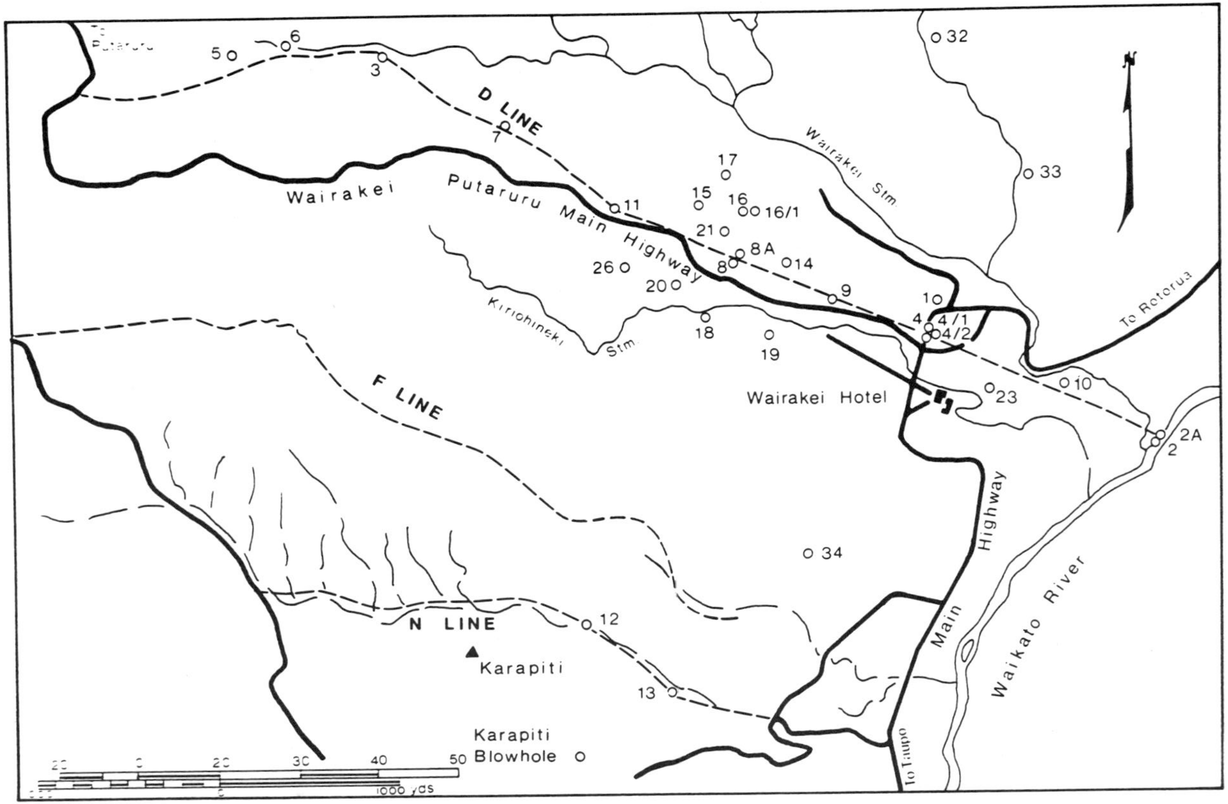

Fig. 2.3. Locations of drill holes.

end of 1950 the encouraging results that had been obtained would be bettered and that more concrete data from which to draw conclusions would soon become available.

Larderello

In 1944, the engineer-in-chief of the Ministry of Works and Development requested the New Zealand Defense Department to assign an engineering officer serving in the armed forces in Italy to investigate the work on geothermal energy that had been undertaken there at Larderello. The idea was to assist the DSIR in gathering more information about the potential of thermal steam being applied to power generation in New Zealand. The Commissioner of Works and the director of the geological survey made a visit to Larderello, Italy in 1948 to study developments there.

They found that the geological conditions at Wairakei differed substantially from those at Larderello. At Larderello, the geothermal steam was trapped under an impermeable cap rock. The steam produced was dry and often had a substantial degree of superheat. At Wairakei, drilling took place through beds of permeable volcanic material where temperatures approached boiling point, thus adding substantially to the difficulties of drilling. While the observations at Larderello were important in providing an understanding of the overall approach to harnessing geothermal power, conditions were so different that most of the work done at Wairakei depended heavily on decisions reached in New Zealand. Perhaps the most important thing Larderello contributed to New Zealand was its success, because there is no doubt that the successful development at Larderello strongly influenced the decision to proceed at Wairakei.

Following the visit of the departmental officials to Larderello, a top New Zealand scientist went there in April 1950. When he returned, he explained to the New Zealand group some of the work which was being done there. He tried to draw recommendations from what he had seen to help New Zealand in its plan to harness geothermal energy. He reported that, as far as he could determine, no magmatic or gravimetric work had been done, nor had any refraction shooting. He also commented on what was later to become important to New Zealand in regard to its tourist industry: that the taking of steam from deep beneath the ground surface reduces or partially eliminates surface thermal activity. General depths of the wells at Larderello were on the order of 1,640 feet (499 meters), though some were only 656 feet (199 meters) deep. The size of drill holes was about 18 inches (46 centimeters) in diameter; one hole was already drilled 1,640 feet (499 meters) deep and arrangements were made to go as deep as 3,281 feet (1,000 meters), if necessary.

The scientist observed that the hot spring area of the Larderello district was only about 7 miles (11.2 kilometers) by 14 miles (22.4 kilometers) and that he personally hoped that the potentialities in New Zealand were far greater than at Larderello. His recommendations about what should be done were made clear when he said: "I have a feeling that the sooner we get a hole of even four or six inches (10 or 15 centimeters) in diameter down to 1,500 feet (457 meters) the better." By doing this, he believed, a truer sample of magmatic gas could be obtained to act as a guide in the selection of material for turbines. One or more of the gases comprising magmatic gas could adversely affect some, but not all, of the materials chosen for the turbines. "I don't think it matters much where you put the first bores," he said, "so long as there are surface indications of steam. If the resulting steam is troublesome it could be easily piped for heat-giving purposes."

The scientist qualified his position by stating: "This does not mean you should not take all the scientific observations you can with regard to the future. But time marches on and you should be prepared to take risks. I expect corrosion problems in the pipes would occur in areas near the surface. A site for the first hole might well be chosen accordingly."(4)

Upon hearing the eminent scientist's advice, specialists from the Geological Survey Office reacted violently. They took a dim view of recommendations about where the first bore should be put and advice to go ahead without much more geological or geophysical study. A New Zealand scientist in the geological survey was concerned that the remarks would do harm among the uninitiated who could take them on the authority of the author rather than on their value. The geological scientist posed the question: "How would anyone be so irresponsible as to say that it does not matter much where the first holes are drilled as long as there are surface indications of steam?"(5)

While good information could be obtained from other projects on the other side of the world there were several handicaps, including the difference in the geological structure of the two areas and a very severe language problem, which prevented a full interchange of ideas. Perhaps the most pressing problem of all at this stage of the project was the pressure from the government to get under way in New Zealand as soon as possible in the face of insufficient data. In this sense, the eminent New Zealand scientist personified the government's position.

In addition to the information received from Larderello, information was also obtained from Iceland where hot water was being obtained from 45 wells of 4 to 8 inches (10 to 20 centimeters) diameter and from 450 feet to 2,400 feet (137 to 731 meters) deep, producing 1,125 tons per hour at a mean temperature of 188.6°F (87°C). A total of 90,000 feet (27 kilometers) of hot water wells had been drilled in Iceland, while at Larderello there were over 140 active holes totalling 98,500 feet (30 kilometers) with an overall capacity of 2,000 tons of steam per hour varying in temperature from 284°F (40°C) to 419°F (215°C) with a static pressure ranging from 5 to 27 atmospheres.

Geochemical investigation

The reports on the chemical aspects of the geothermal research in New Zealand resulted in tentative deductions. The sodium potassium ratio of water from hot springs agreed with analyses from the other parts of the world, the values being low for active acid thermal areas and high for cold weak areas, or areas with deep water circulation. Wairakei appeared at that time to be an area with deep water circulation. The experts felt, however, that analyses of other areas were necessary, and they strongly recommended that chemical investigations should seek further constituents of magmatic origin. A later geothermal chemical report said that the gases discharged from the drill holes contained, for the most part, constituents normally expected from thermal spring gases, but the amount of gas was relatively lower than was usual in fumaroles and was low in comparison with the gas content of Larderello steam. The inference drawn from the low gas emission was that the magmatic steam was probably diluted by ground water. But the temperatures recorded in the holes in relation to the amount of dilution of the steam seemed to indicate that the magmatic steam was very hot before dilution took place.

The geochemical report also indicated that from the results so far obtained on gas content, bores 1 and 4 were located nearest to the most active area (see fig. 2.3). This contention was supported by the evidence of high hydrogen sulfide content. The report said that the amount of boric acid in the steam from the Wairakei holes was well below the amount found in the steam at Larderello. The report suggested that corrosion of metals might be expected to occur to a greater extent at Wairakei than at Larderello due to the presence of water-containing chloride in association with the high hydrogen sulfide content. There was no water at Larderello. The report suggested the use of stainless steel as a possible remedy. At this stage, however, conclusions drawn from the geochemical reports were only tentative.

Steam output

The next major area reported on was the measurement of steam output from drill holes. Measurements of steam and heat from the drill holes were taken with the use of calorimeters which were made by the staff engaged in investigation work. Calorimeter tests were made on holes 1, 4, 6, and 7. A total of 31 tests was made, requiring 138 runs of the calorimeter. Table 2.1 summarizes the figures given at that time.

The conclusions drawn from these data were that drill holes 4 and 1 produced the greatest amounts of steam and drill hole 6 produced the greatest quantity of steam and water, probably as a result of the greater diameter of the drill hole. On the basis of 25 pounds (11 kilograms) of steam per kWh and that the production of steam from a 4-inch (10 centimeter) drill hole was 2.5 tons per hour, the power output from such a well would be 224 kW and it would produce 1.96×10^6 kWh/year. The heat output of a 4-inch (10-centimeter) hole at 7.5×10^6

Table 2.1. Summary of Test Results

Hole number	Hole diameter – inches (centimeters)	Depth – feet (meters)	Casing – feet (meters)	Vapor pressure – lbs per square inch	Number of runs of calorimeter	Steam and water discharge	Heat – Btu per hour	Steam – percentage	Steam discharge – tons per hour
1	4 (10)	600 (183)	470 (143)	-	66	7.38	7.56×10^6	32.2	2.38
4	4 (10)	600 (183)	487 (148)	15-26	51	2.64	6.62×10^6	89.1	2.35
4	4 (10)	910 (277)	882 (269)	18	4	7.42	9.3×10^6	39.3	2.91
4	4 (10)	910 (277)	882 (269)	25	2	6.37	8.05×10^6	39.8	2.53
4	4 (10)	910 (277)	882 (269)	30	2	5.85	6.98×10^6	35.9	2.10
4	4 (10)	910 (277)	882 (269)	36	2	5.39	6.51×10^6	37.0	1.99
6	5½ (14)	429 (131)	128 (39)	50	1	13.15	12.5×10^6	25.0	3.29
6	5½ (14)	429 (131)	128 (39)	25	1	5.35	5.0×10^6	32.0	1.71
6	5½ (14)	429 (131)	128 (39)	38	1	8.04	9.42×10^6	34.0	2.73
6	5½ (14)	429 (131)	128 (39)	48	1	6.20	6.87×10^6	32.0	1.98
7	4 (10)	550 (168)	475 (145)	7.5	5	0.36	0.9×10^6	83.7	0.30

Source: "Notes on Geothermal Investigation at Wairakei, New Zealand," Ministry of Works and Development File P.W. 92/14/20.

Btu per hour, assuming an efficiency of 10 percent, would amount to 220 kW or 1.927×10^6 kWh per year.

Surface temperature gradients

The rate of heat loss by conduction was assessed at 10,000 Btu per hour per acre. This rate of heat loss would be 6.4×10^6 Btu per hour per square mile, which roughly equalled the heat output from a 4-inch (10-centimeter) diameter drill hole in Wairakei. At 10 percent efficiency, the power output from 1 square mile (2.56 square kilometers) in the Wairakei area would be a minimum of about 200 kWh.

Ground magnetic surveys in the Wairakei-Taupo area

Surface magnetic surveys were undertaken in the Wairakei area, around Lake Rotokawa, and in the Spa Hotel vicinity. The value of a magnetic survey depends on the occurrence of low vertical magnetic field intensity in areas where thermal activity has altered the nature of the ground and has caused reduction of the magnetic properties of the rock. Magnetic investigations were not conclusive, however, as there were other possible causes of low magnetic intensity. Therefore, magnetic work had to be evaluated in conjunction with all available evidence before deductions could be made. It was possible that the detection of thermal-altered areas by magnetic means might reveal areas in which there was presently no surface manifestation of thermal activity.

Work on magnetic susceptibility revealed that core samples from holes 4 and 2A had generally higher susceptibilities than those from holes 5 and 3. The difference was probably due to the differing degrees of alteration of the magnetic minerals, but might equally have been a result of the presence of iron sulfide believed to have been found in holes 4 and 2A. The inference drawn from this thermal alteration theory was that the area at the east end of the "D" line at holes 4 and 2A was more thermally active than was the area at the west end of the "D" line at holes 5 and 3 (see fig. 2.4).

Geology of the Wairakei area

The main geological features of Wairakei were determined by an intense though rushed investigation. The study showed that an anticline, trending north-northwest with a fault running parallel to its axis on the eastern side, extended from near the Karapiti fumarole to beyond the "D" line near hole number 3 (see fig. 2.3).

An additional fault extended from this fault and struck north-northeast to the Wairakei Valley. The uppermost formation in the area was recent alluvium varying in thickness from 50 feet (15 meters) to 100 feet (30 meters) and consisting of pumice, gravels, and ash washed into the valleys. The alluvium allowed ready movement of ground water. The next lower formation, called the Wairakei formation, consisted of soft, massive, ill-sorted, unbedded breccia about 200 feet

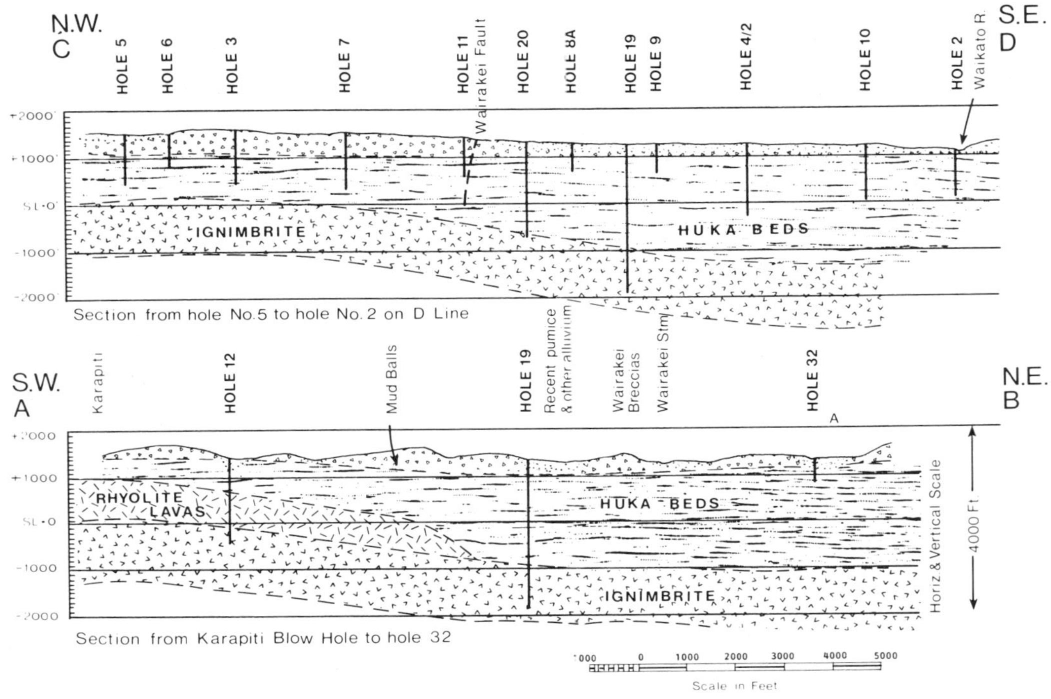

Fig. 2.4. Cross section of "D" line.

(61 meters) thick, most of which had been removed from the Wairakei area by erosion. Beneath it lay a roughly bedded fine breccia varying in thickness from 50 feet (76 meters) to 500 feet (152 meters). Underneath the breccia lay well-bedded pumiceous sediments, probably lacustrine in origin, ranging from fine conglomerates to mudstones. Seismic prospecting suggested that these beds extended down to several thousand feet. Geologists felt that the Wairakei formation was a remnant of an old sheet of sediments which had undergone folding and faulting movements which allowed the passage of steam from deep within the earth to cause the thermal activity. They found that the fold structure of the anticline was transversed by a fault which must have extended quite deeply if it was the cause of the thermal activity. The arrangement of the bedding in the lowest formation, with its various lenses of material of various textures, probably provided traps and passages for steam at different levels in the sediments.

Based on the information available, investigators decided that drilling for steam should be carried into the lowest formation, the Huka Beds. The depth formation of these beds at Wairakei varied from 85 to 450 feet (26 to 137 meters). More drilling was obviously needed to verify geological hypotheses and to test the tentative conclusions already reached.

Geological and geophysical drill: log data

A study of the temperature gradients in drill holes 1 to 8 on the "D" line showed that the ground temperature rose relatively slowly with depth until the Huka Beds were reached and the maximum temperature occurred generally after the Huka Beds were penetrated 100 feet to 200 feet (30 meters to 61 meters). The maximum temperature occurred 550 to 600 feet (168 to 183 meters) below the ground surface. Variations did occur, though, so these figures were not constant. And in some holes, temperatures would obviously increase if the holes were drilled deeper. Investigators found that the temperature reached a maximum in some holes, and then decreased. With the exception of holes 2A and 5, the boiling point for depth relationship showed that the boiling point occurred approximately 550 feet (168 meters) down. The boiling point for depth relationship is a relationship between depth and the temperature at which water will boil, taking into account the variation in the density of water with temperature. Cold ground water appeared to be entering hole 2A at about 300 feet (91 meters) due to insufficient casing, while hole 5 appeared to be far from the source of geothermal heat. The source of the geothermal heat was the hot, circulating ground water.

Of interest to investigators at this time was the fact that holes 2A and 5 were at opposite ends of the "D" line, both were the deepest holes drilled so far – 1,000 feet (305 meters) – and both had the least amount of casing. Hole 2A consisted of 186 feet (57 meters) of 4-inch (10-centimeter) diameter casing and hole 5 consisted of 380 feet (116 meters) of 3½-inch (8.9-centimeter) diameter casing. With the exception of hole 8, which was still being drilled at that stage and had not had any 4-inch (10-centimeter) diameter casing placed in it, all the other drill holes had at least 460 feet (140 meters) of 4-inch (10-centimeter) casing. The low boiling point for depth relationship in holes 2A and 5 was probably due to the entry of cold ground water into the holes because they were not cased deeply enough. If these holes had been cased to 460 feet (140 meters) as had the others, it is probable that the boiling point for depth relationship would have shown that boiling was occurring at a much higher level.

Geological report on drilled holes

Preliminary findings were based on the geological data obtained from eleven holes drilled for steam at Wairakei, totaling 6,273 feet (1,912 meters). Thirty-nine other holes, totaling 7,430 feet (2,265 meters), had been drilled for seismic survey work. The seismic holes were shallow, reaching a maximum depth of only 1,020 feet (311 meters). The holes for steam were mostly core drilled, while the drill cuttings from the seismic survey holes – little bits of debris that came to the surface as they were drilled – were examined by the geologists.

From an examination of the drilled cores and washings, along with the knowledge of the makeup of the surface geology, the rocks and formations were classified. Faults were found parallel and perpendicular to the bedding. Hydrothermally altered pumice sandstones and siltstone imbedded with mudstone were also found. Silicified bands occurred at various levels throughout the Huka Beds.

Temperature measurements in the seismic survey holes showed that thermal gradients were generally steeper where the Huka Beds were drilled. Higher temperatures were found in these beds than in the unfractured Wairakei formations or in the pumice alluvial cover.

Investigations confirmed that the Huka Beds were sedimentary in origin, consisting of interstratified beds of fine and coarse materials, fractured by faults, and containing the highest temperatures. Hydrothermal alteration and silification of the beds were also confirmed, so that the geological conditions were satisfactory for the passage of steam upward from deep below and for the passage of steam horizontally along beds either by the horizontal fault planes or through pervious beds.

Petrological investigations of cores

This facet of the investigation for steam at Wairakei involved a great deal of work and thought by the petrologists of the New Zealand Geological Survey and added considerably to the knowledge of the characteristics of the rocks through which the drill holes had passed.

Tentative correlation of rock types between drill holes suggested faulting, which was confirmed by the surface geological survey. Similar sequences of beds in drill holes suggested also that there was a slight dip from east to west of $1^{\circ}40'$.

Alteration of rocks in varying degrees was noted in the cores, and there was a general increase of alteration with depth. The presence of secondary minerals formed by new substances brought in by ascending solutions and vapors revealed that the metamorphic processes were of a metasomatic nature.

The cores showed varying degrees of porosity. The pores were not always continuous and were often of microscopic size only. Pumice was the most permeable constituent. The petrographic evidence suggested that the steam ascended through the pores of the rocks and not through fissures, or large spaces. The belief was that both hydrothermal solutions and vapors were active at Wairakei.

This information, together with the small vertical range of alunite found in hole 6, raised the question of how much of the ground water was of meteoric origin and how much was of magmatic origin. If the ground water was mainly of magmatic origin, the removal of a substantial quantity of it by jetting from the active drill holes might bring dry steam much closer to the permeable beds in the Huka formation.

Investigators decided that active drill holes should be made to discharge continuously while ground water measurements of surface level were made from adjacent shallow holes put down for this purpose. The evidence obtained up until that time presented a fairly clear picture of the conditions at shallow depth. Better results could have been obtained by drilling along the fractures, but larger quantities of steam were likely to be obtained only from wells drilled to greater depths.

The evidence then available confirmed the interstratification of permeable and impermeable beds and also confirmed the existence of fault zones which contained steam and fed these beds. This evidence suggested a very good arrangement for obtaining steam and hot water, with the probability that steam without hot water might be tapped at greater depths. This information contributed invaluably to the investigation and represented a great step forward in the knowledge of the geological conditions likely to be encountered in the graben extending from Taupo to the Bay of Plenty coastline.

Airborne magnetometer survey

The first airborne magnetic survey in New Zealand took place in conjunction with the Wairakei project. The work was carried out by the Royal New Zealand Air Force (RNZAF) in cooperation with the Magnetic Survey Branch of the DSIR. This survey was successful in that it confirmed the results of the ground magnetic survey carried out over some of the same area. The airborne magnetic survey covered the entire 13-by-13 mile (20-by-20 kilometer) area of Taupo and Wairakei, which was a much greater area than could be covered by the ground parties in the time available for producing a magnetic survey and a report by the end of 1950.

Seismic survey

DSIR geophysicists concentrated on seismic refraction work in the Wairakei area to obtain surface and near-surface velocities of sound, or seismic waves in the formation. They also did seismic reflection work to see if high-speed refractors existed deep below ground surface. High-speed refractors are geological formations that are normally hard and dense. Velocities through the various near-surface formations were obtained, and geophysicists determined that no high-velocity layer existed within 1,000 feet (305 meters) of the surface. This meant that no hard dense formations existed within 1,000 feet (305 meters).

The reflections obtained from the seismic reflection survey were poor, making the interpretation of the records very difficult. From velocity data obtained, the average normal velocity increased at first rather rapidly and then gradually decreased. Further investigation was necessary, however, before any sound conclusions could be reached.

Another item of interest was that on the assumption of an average velocity of 6,000 feet (1,829 meters) per second, a depth of approximately 6,000 feet (1,829 meters) was obtained for a horizon at which the reflections were more or less continuous across the whole profile section. Deeper reflections than this had also been recorded. This geophysical interpretation suggested that a hard formation existed at 6,000 feet (1,829 meters) which extended over the entire section measured.

The results of the seismic reflection survey on "F" line indicated the presence of an anticlinical structure at depth.

Gravity survey

A gravity survey revealed the presence of two major features: a change of basement level of at least 5,000 feet (1,524 meters) occurring at the eastern edge of the graben, and a pronounced gravity "high," striking northeast in the Wairakei area. It was considered not advisable at this stage to put the gravity results on a quantitative basis until the magnetic and seismic surveys had been correlated with the gravity results.

It was evident, however, that most of the thermal activity was associated with subsidiary features such as cross faulting, although the possibility of faulting on the flanks of the main structure could not be ruled out until the data had been placed on a quantitative basis.

Heat source

A report by the Department of Scientific and Industrial Research gave an assessment of how much geothermal energy existed in the Wairakei area up to a depth of 1,000 feet (305 meters). It assessed the amount of energy to be a minimum of 600,000 kW and a maximum of 4,000,000 kW based on the evidence available at that time. On the basis of 15 percent overall efficiency, this would give an output of electrical power between 90,000 kW and 600,000 kW from the Wairakei area if all the heat could be harnessed.

Ministry of Works and Development activities

The camp at Taupo township had been built by the MWD to accommodate the staff of the various government departments engaged in the investigation work at Wairakei, together with service buildings for offices, stores, and laboratory.

Survey work included 113 miles (182 kilometers) of leveling and completed the requirements of the DSIR by the end of 1950. Plans were prepared for the Wairakei-Taupo area, showing the various traverse lines surveyed and leveled for the DSIR.

The drilling of prospecting holes for steam was carried out by four Ministry of Works and Development drills, two drills supplied by DSIR complete with crews, and one large drill supplied by the Mines Department complete with its crew. The Ministry of Works and Development drills were assembled from various parts of New Zealand, and crews had to be recruited and trained as the work proceeded, which in itself reflected credit on the drilling superintendent and staff. Explosion-preventer gear for the protection of the workmen and drilling equipment had to be designed and built, and a calorimeter was designed and built for measurement of the discharge from the drill holes.

Although the original intention at the start of the investigation was to drill holes down to 500 feet (152 meters) with one deep hole extending to 1,000 feet (305 meters), most of the holes were drilled to 600 feet (183 meters) and three were drilled to at least 1,000 feet (305 meters). This was because the drilling work progressed more rapidly than expected. A total of 8,153 feet (2,485 meters) of drilling had been completed by the end of 1950.

Some considerable effort and time was spent on drilling hole 4 down to the limit of the equipment available. At one point, it became impossible to proceed any deeper because the casing was damaged. Several unusual features cropped up in the behavior of this deep hole during the drilling. The static pressure at the surface in the 4-inch (10-centimeter) casing averaged 90 pounds (41 kilograms) per square inch (6.25 square centimeters) with little variation. If the valve on the 4-inch (10-centimeter) casing was opened to any extent while water was not being pumped down the hole, steam appeared at the surface in less than 30 seconds. While lowering or withdrawing the drilling rods from the hole, water had to be pumped down the 4-inch (10-centimeter) casing to cool these rods and prevent the water in the rods from being boiled out each time a length of drill rod was unscrewed.

Besides drilling work, the construction of several miles of access roads to get the drills to the required sites, the construction of cold water pipelines and reservoirs, and the construction of the concrete lined pits for the blowout-preventer equipment were a substantial part of the work undertaken by the Ministry of Works and Development.

Prefeasibility and Appraisal

Preliminary assessment

By the beginning of 1951, it was evident that there were three principal questions which needed to be answered in regard to geothermal development at Wairakei. These were: 1) What was the continuous power output of the area? 2) How much of this power could be harnessed? 3) How much would it cost? Although a good deal of investigation had been carried out, those responsible for the project were unable to answer these essential questions.

Certain factors, however, were clearly established. It was clear that the origin of the heat was magmatic and that the heat escaped to the surface through faults, fissures, and pervious strata, heating the ground and ground water in its passage. The heat was eventually given off by radiation from the surface, by conduction, and by convection currents in the underground water. Investigators felt that to a depth of at least 1,500 (457) and at most 6,000 feet (1,829 meters), the geological structure comprised sedimentary and volcanic beds which were pervious and fractured by faults. Some of the beds, they speculated, might be gently folded, in which case the steam rising from the magma could be trapped in a dome, but at this stage there was no conclusive evidence that a dome existed. On the other hand, its existence had not been disproved, either. There was some dispute over what the significance of a dome area would be. The geologists suggested that only steam trapped in a dome would be worth exploiting for power purposes, and if one could not be found at Wairakei, development should not be undertaken there. The engineers, however, felt that it was a matter of opinion whether a dome was necessary to proceed. They thought that it would be possible to obtain all the power needed for many years without a dome, provided the steam could reach the production wells by means of fissures and porous strata.

What had been established was that it would be possible to get approximately 300 kW from a 4-inch (10-centimeter) diameter well 600 feet (183 meters) down. Investigators speculated that it should be possible to get four or five times as much power from an 8-inch (20-centimeter) diameter well of the same depth if the steam supply was in fact relatively shallow and, consequently, each well could be drilled at a reasonable cost. It appeared that if there was a dome in the Wairakei area, it would be at considerable depth and the cost of tapping it would be high.

A number of unknown factors still puzzled the investigating team. Would a large number of production wells exhaust the steam supply, and if so, how many wells could be sunk before it became unprofitable to continue drilling? What about the interaction between wells? The answer to this question would also affect the number of wells that could be drilled. And finally, would the drilling of deep wells decrease the yield of shallow wells? Opinions differed on this issue. One side said yes, it would decrease the shallow wells' yield; the other side said that this result was unlikely because the strata of the Wairakei and Huka formations were so porous and fractured that a few extra channels provided for the passage of steam by deep wells would be insignificant compared with existing natural channels. Similarly, the yield of wells remote from faults might be as large as or even larger than the wells near faults.

Investigators decided that as the cost of drilling deep wells was great, they should not develop wells deeper than 1,000 feet (305 meters) and that future investigations should be made to ascertain how much power was available up to this depth. As the geophysical work already carried out had shown that there was no significant change in the geological structure for several thousand feet, investigators believed that there was little to gain in drilling beyond 1,000 feet (305 meters), even if the drilling equipment could go this deep. The investigating engineer recommended an immediate program to complete the prospecting holes in "D" line at Wairakei (fig. 2.3) and at Karapiti, as work had been done on those already. His investigating team believed that all holes should be blown and kept under observation for several months, possibly a year. They believed the most important work to be drilling and testing of development wells at bore hold 4. Until this was done, they felt, no other prospecting should be undertaken.

Need for urgency

While investigation work continued in 1952 under the DSIR and the Ministry of Works and Development, the minister of works and development told the MWD that his government had decided that there was an urgent need to prove within 12 months the existence of a 20,000-kW geothermal steam power source at Wairakei. The investigation program had already been condensed from what was regarded as a tight program, and now the government was pressing harder to get information and quick results to ease New Zealand's energy crunch. With this in mind, the Geothermal Advisory Committee set up a subcommittee consisting of an officer each from the Ministry of Works and Development, the State Hydro-Electric Department, and the New Zealand Geological Survey Office. They were asked to come up with a plan detailing a course of action to meet the new 12-month deadline. The subcommittee reported that the existence of geothermal steam to develop 20,000 kW could only be proved by actually producing that amount from drill holes at Wairakei and allowing it to discharge over a period of time to assess the rate of decrease of discharge, if any. The subcommittee felt that if 20,000 kW could not be proved to exist within the 12-month deadline, the amount of steam they were able to find could eventually be developed economically.

The subcommittee agreed that the DSIR should fix the location of 16 drill holes at Wairakei in positions where sufficient steam would most likely be available. Although 16 holes might be more than could be drilled in the 12 months, the subcommittee still wanted enough holes drilled to be sure of getting 20,000 kW eventually. They agreed that the location of some holes might have to be altered as the drilling proceeded.

The number of holes was based on 20,000 kW, requiring 20 pounds (9 kilograms of steam per hour per kW for condensing units, with a dryness factor of 25 percent. This would require 1,600,000 pounds (720,000 kilograms) of steam and water per hour of 100,000 pounds (45,000 kilograms) per hour of the mixture from each of the 16 holes. The Ministry of Works and Development was asked what drilling performance could be obtained by the available drilling equipment together with additional equipment to be ordered. The subcommittee decided that the drilling personnel should be increased to three shifts per day, six days per week to cope with the major work. It was necessary to consider the distance between wells, and they felt that 600-feet (183-meter) centers would provide ample clearance from each other but this would have to be determined after discussion with the DSIR.

The subcommittee felt that to determine if 20,000 kW was available, it would be necessary to allow all holes to discharge as long as possible, including those holes already drilled along the "D" line. They suggested that separators, standard nozzles, and gauges be provided to record changes in temperature and pressure on as many holes as possible in order to assist in the assessment of the power availability.

The subcommittee agreed that the method of drilling, casing, and grouting the holes might have to be modified considerably when required for permanent development. Engineers would have to pay close attention to the type and diameter of casing, to grouting both in the earth and around the casing, and to the depth of uncased holes below the permanent casing, in order to obtain the maximum results both for the life of the well and for the amount of production in relation to the amount of drilling.

Organizational change

The new 12-month deadline to prove that 20,000 kW of geothermal steam could be drawn at Wairakei necessitated organizational changes. There had been confusion in the past over who in the chain of command was responsible for what. One of the reasons for this confusion was that the Ministry of Works and Development was not experienced in working under the direction of a committee.

The Geothermal Advisory Committee, however, acting in its capacity of policy maker, had decided to drill for steam to develop 20,000 kW. The GAC had then asked that this be done by the MWD. The engineer-in-chief of the MWD was instructed accordingly, and approval was given for the work to be dealt with as a project in the normal manner with all technical direction emanating from the engineer-in-chief of the MWD. The work did, however, differ from normal work in one main respect, and that was that rather than proceeding on a department-to-department basis, it had been decided that matters normally dealt with in this manner be dealt with formally by an advisory liaison committee. This deviation from the norm, however, in no way affected the instruction that Wairakei be considered and dealt with as a normal project.

At a meeting in June 1952, the Geothermal Advisory Committee came to a number of important decisions regarding the future administration of the Wairakei Region. These included:

1. The revocation of an earlier decision to place the Wairakei project in the hands of the SHED and make it the responsibility of the Ministry of Works and Development.
2. The disbanding of the committee due to the change in incidence of the work at Wairakei from investigation to development.
3. Following disbanding, the handling of policy and other major matters by meetings between the heads of the MWD, the SHED, and the DSIR.
4. The setting up by the three departments of a liaison committee, with the MWD representative as convener.

The Geothermal Advisory Committee was accordingly disbanded, and the administration of the Wairakei project proceeded as outlined at their last meeting.

The establishment of a liaison committee of departmental officers recognized that the nature of the work at Wairakei still required much collaboration among the three departments. This committee was, however, purely advisory. Executive action was to be the function of the existing departmental organizations. The liaison committee would be required to keep details of the work under constant review, make any recommendations necessary, and give advice on matters referred to it by the three departments. The personnel of the committee were to be the chief electrical engineer of the SHED, the director of the geological survey, and the assistant engineer-in-chief of the Ministry of Works and Development. The latter was to be the convener of the committee, and all directions and instructions on work at Wairakei were to be his responsibility.

Over the years, organizational difficulties became apparent. In retrospect, it is clear that this was due to the continually changing nature of the work, from purely scientific in the early days to engineering later. There was no clear transition point, but by 1952 the Geothermal Advisory Committee recognized the problems and placed the work in the hands of the MWD. The unusual nature of the work was also recognized by the establishment of an interdepartmental advisory committee. The basic difference between this interdepartmental advisory committee and the former Geothermal Advisory Committee was that the GAC was the responsibility of the DSIR and established by cabinet direction, whereas the interdepartmental advisory committee was responsible to the assistant engineer-in-chief of the MWD and was established by agreement among officials of the government departments concerned with the project – namely, the MWD, the DSIR, and the SHED.

Approval in principle of combined project

Following the establishment of Wairakei as a project in late 1949, drilling proceeded rapidly so that by early 1953, 20 MW had been established at Wairakei by shallow drilling. Engineers from the MWD agreed that two deep wells should be drilled, using drilling rigs which had recently been purchased. Even without these deep wells, results were considered to be so encouraging that the New Zealand Cabinet, at a meeting in May 1953, took the following actions:

1. Approved in principle the construction of a heavy water and electricity generating plant using geothermal steam subject to these conditions:
   a. Availability of sufficient steam,
   b. Consultants' report acceptable to the government, and
   c. Satisfactory financial arrangements made with the United Kingdom government.
2. Authorized the MWD to carry out further investigations under a £200,000 budget.
3. Authorized the DSIR to carry out corrosion and other tests at an estimated cost of £9,100.
4. Authorized the Minister of Works and Development to hire consultants.

Some time previously, the United Kingdom Atomic Energy Authority (UKAEA) had expressed interest in the possibility of producing heavy water using geothermal energy at Wairakei. Heavy water was an essential ingredient needed by the atomic energy industry. The proposal had been discussed by the New Zealand prime minister and United Kingdom officials during a visit to England in 1952, and again with the director of the UKAEA establishment at Harwell, England, when he visited New Zealand later in the same year. These discussions, together with the promising nature of the energy source, lead eventually to the approval in principle of the construction of a plant which would produce geothermal energy to help meet New Zealand's power needs and provide heavy water for the UKAEA.

Feasibility

By the beginning of 1954, the investigation work and the drilling had proceeded as quickly as could be expected considering the available equipment – much of which was not commercially available and had to be made by local engineers – and inexperience of some of the personnel on the project. There was also a budgetary constraint. The Wairakei geothermal investigation expenditure for 1953-1954 had a ceiling of NZ$800,000, and remaining beneath this ceiling was proving difficult. Assistance was given by providing resources such as housing and other facilities out of the general budget of the MWD.

The investigation work at Wairakei had been substantially finished by the end of 1954. It was now a matter of drilling for power when it was known how much was required. In the opinion of the assistant engineer-in-chief of the MWD, the success of the plan to use geothermal steam to generate electric power could only be determined by undertaking a series of trials. He felt that all the evidence pointed to the fact that geothermal steam had a legitimate place somewhere between cheap hydro power and more expensive coal-fired steam power. This conclusion therefore warranted the expansion of geothermal power ahead of some of the more costly hydroelectric development which had already been proposed by the government. The ultimate limit of development of the geothermal resource would be either the constraint of power distribution or the availability of geothermal heat.

As far as available power went, there were two conflicting theories. The assistant engineer-in-chief of the MWD took a different view from that of the DSIR. The DSIR measured the heat outflow from the surface manifestation of the geothermal heat. They figured that the heat input and the heat output would reach a balance through the years, and therefore heat output could be taken as the minimum power available for development. The assistant engineer-in-chief of the MWD said that the heat output was a function of the conductivity of the earth's crust rather than of the heat available, and if the earth's crust was partially bypassed by drilling to a great depth, the heat output, and therefore the power available, would certainly exceed what could be obtained at the earth's surface.

Both felt that the extent of the geothermal heat could be determined only by drilling holes until a new balance of heat input and output was reached. The drilling would have to progress at a not too rapid rate because the balance could not be easily discerned. This was because it was masked by the large amount of stored heat available. If one assumes a temperature drop of 212°F (100°C), then the power available from this stored heat in one cubic mile (four cubic kilometers) of land could produce 100,000 kW for 30 years.

The assistant engineer-in-chief considered that the development of a geothermal field for power would require something on the order of ten years. On this basis, it was necessary that investigation should begin as soon as possible on new fields. The MWD felt that it would be prudent to extend preliminary investigations to new fields and continue development drilling at Wairakei at the rate of three holes a year because the energy problem in New Zealand was still critical.

The Ministry of Works and Development was anxious to have a firm statement on what actually was required at Wairakei. That department wanted to take the matter up with the prime minister, who would most likely want to be involved in the decision-making process of such an important project.

Engagement of consultants

The need for assistance from consultants arose mainly from a general shortage of staff, and in particular a shortage of staff skilled in steam power station work. New Zealand's lack of experience in this area made the selection of suitable consultants difficult. The presidents of the Institution of Civil Engineers and the Institution of Mechanical Engineers were therefore asked for advice. From their suggestions a selection was made, and in July 1953 a firm of consultants from England was asked by the MWD if they could be interested in undertaking the following duties:

1. Prepare a report on the use of geothermal steam.
2. Prepare a report on the production of heavy water.
3. Prepare a report and cost estimate for a power station initially capable of producing 20,000 kW.
4. Provide general advisory services.
5. Provide services in connection with specific projects as required.

The English consultants accepted, and their appointment was confirmed in August 1953. It should be noted that although the New Zealand government proposed cooperating with the United Kingdom government in the production of heavy water, the primary function of these consultants was to report on the establishment of a 20,000-kW geothermal power station. The heavy water portion was secondary.

Consultants' reports

On confirmation of their appointment, the consultants immediately began work on the stipulated report. Their chief development engineer visited New Zealand to gather pertinent information on the possibility of developing geothermal steam, and to advise the New Zealand investigators what further information would be needed to prepare a preliminary report and cost estimate for a 20,000-kW geothermal power station.

They were also required to comment on proposals being prepared by the consultants to the UKAEA on the production of heavy water. In September 1953, however, it became clear that a heavy water reactor would not be best suited to meet the civil and military requirements of the United Kingdom. Although the UKAEA was no longer prepared to support the project financially, they were very ready to give technical support to the New Zealand government if it had wished to continue with the heavy water project. The New Zealand government did not proceed with the project, however, and the consultants were not required to report on heavy water at that time.

On June 22, 1954, the English consultants sent to the minister of Works and Development a special progress statement outlining what had been done so far and spelling out the position as they saw it at that date. They were reluctant to prepare a formal report until they had time to consider information on well outputs they had just received.

The points considered in the progress statement on June 22, 1954 were as follows:

1. The power potential of steam to be produced from bores.
2. The extent to which plans would be based only on proven steam. Should they anticipate the results from later bores, which would affect the size of the initial and future development of the power station?
3. Whether the corrosive nature of the steam would be such as to need a further, long period of investigation before designs could be prepared, or, at worst, make development for power purposes impractical or uneconomical.
4. Whether the station should be situated near the bores or near the Waikato River, where cooling water would be directly available.
5. The method of separating the water from the steam issuing from the bores.
6. The best method of dealing with the different pressure characteristics of the deep and shallow wells.

Technical aspects

Water: steam ratio. The ratio of water to steam issuing from well 20 was at least 2:1. It was necessary to use a system of separation to remove the water from the steam for use in the turbines. A simple form of mechanical separation was in use and the English consultants found it to be effective. As time passed, however, further development of a practical separator suitable for the operation of a power station became necessary. The main need for effectively separating the water and the steam was to get rid of the harmful salt which was dissolved in the water and which, unless kept to a minute quantity, would accelerate corrosion of the turbines. The plan at this stage was to have a system of double separation which would keep the salt quantity down to a tolerable limit.

Corrosion. The steam contained chemicals – in particular, hydrogen sulfide, which was a highly corrosive substance, particularly in the presence of oxygen. Both the English consultants and the MWD thought that the corrosion problem could in fact be overcome partly in the design stage and partly by operating procedures which could be laid down later. They took hope from the fact that the corrosion problem at Larderello had been overcome. The consultants knew, however, that it was only by operation on a full scale that truly useful results could be obtained since it was difficult to design satisfactory small-scale tests which would adequately simulate the full-scale conditions.

Yield of wells. Consultants then proceeded on the assumption that there would be a number of deep bores giving a yield of steam on the order of 200,000 pounds per hour. Each bore would be good for a power output of some 14,000 kW, and on this basis they envisaged a station of some

50,000 kW. Later information, however, showed that the bores that they had anticipated to be good for 14,000 kW were in fact only good for about 6,000 kW. The shallow bores already drilled, however, collectively yielded sufficient steam for approximately 17,000 kW. So they decided that, keeping a little in hand, it would be safe to go ahead with an installation of 26,000 kW and a small high-pressure set of about 2,000 kW. An extension to approximately 50,000 kW was planned when bores 18 and 19 were cleared of obstructions and bore 35 sunk. The consultants considered that a temporary installation of 3,000 kW would be an advantage. This machine would discharge into the atmosphere using the steam from bore 20. They felt that this process would provide advance running experience and assumed that its output could be used by feeding the steam into a local distribution line, perhaps to the Wairakei Hotel.

Site of station. The consultants decided that it would be better to transport the steam to the powerhouse rather than transport the cooling water to the steam field, and so the station should be situated near the Waikato River. The best site seemed to be a level terrace south of the road to Rotorua. The consultants felt that a conservative estimate for the cost of a power station should be based on a life span of 15 years. But they would not say whether the wells would fall off markedly in yield during that period. They could keep the total yield up by drilling more bores, but it was possible the whole region might be depleted in a comparable time. This loss of steam, however, did not happen at Larderello, a project which was much larger than Wairakei, producing 280,000 kW. The question of corrosion would be another factor in determining the overall life of the plant.

Overall conclusions

The overall conclusions and assumptions of the consultants were that further steam could be found in the vicinity to supply a station of some 50,000 kW. They were satisfied that the geothermal steam was capable of practical development and had every prospect of proving an economical source of power. They felt that a station on the order of 50,000 kW could be constructed for a cost of about $5.5 million and this would lead to a cost of generation of about $10 per annum per kW, or, alternatively, 0.175 cent per kWh. They recommended that the budget for the first stage of 26,000 kW should be about $3.6 million.

Committal

Introduction

In his budget speech on July 22, 1954, the prime minister announced the decision to proceed with the installation of a plant to produce heavy water and electric power as a joint venture of the United Kingdom Atomic Energy Authority and the New Zealand government. The budget, which reviews the government's expenditure for the past year and indicates expenditures for the forthcoming year, is an important policy document. It is used by the New Zealand government as a means of introducing new policies, particularly where these involve substantial expenditure.

In making the announcement, the Prime Minister said:

> Scientific research and engineering investigations have shown that considerable power can be produced economically from this underground steam. At the same time scientists in the United Kingdom have been investigating the possibilities of producing heavy water from the same source. Heavy water is wanted for use in certain types of nuclear reactors for producing atomic energy. The same flow of steam can be used economically to produce both electricity and heavy water. Following negotiations, I am pleased to announce that general agreement has been reached with the United Kingdom Atomic Energy Authority for launching a joint enterprise for the production of electric-power and heavy water in the same plant from the same geothermal steam bores at Wairakei.
>
> The designing of the dual-purpose plant will be undertaken in England, and will require the skill and knowledge of some of the world's leading scientists and power engineers. In this venture New Zealand will be making a worthwhile and unique contribution not only to its own power resources, but also to the development of atomic energy, which, used for peaceful purposes, may well revolutionize the world's industrial processes.
>
> The initial unit to be installed will produce some forty thousand kilowatts of electric-power and further additions to the plant can be made later on in the light of experience gained in this initial plant.
>
> New Zealand is indeed fortunate in having these latent subterranean power resources. We are also fortunate in having scientists, engineers, and workmen here, and in the United Kingdom, capable of designing plant and equipment to convert these roaring jets of steam to use for the benefit of the nation.(6)

A reappraisal of the possibility of using heavy water as a moderator and a coolant in later stages of the United Kingdom power program appears to have led to a reversal of the decision made in the previous year.

Public reaction to the program

The geothermal power project had been promoted fairly extensively by the government as a possibility for improving the power supply in New

Zealand. To the public, however, it seemed a long way from tourist attractions of steam bores and geysers to harnessing such an uncontrollable natural resource. The press seized upon the idea of a heavy water project, emphasizing its economic potential. Indeed, in 1954, heavy water was worth about $140 per pound, so the return from its production would be handsome. The MWD and New Zealand treasury officials also had an eye on this substantial return. Heavy water was a valuable agent in atomic development and was generally a costly item to produce. The need to generate enormous heat to produce it usually involved a great deal of money, but New Zealand's great advantage was that the heat was available naturally. New Zealand officials felt that the country could join a worldwide program to develop heavy water in direct association with Great Britain, thus enhancing their status in the world community. The press proclaimed: "The country has had nothing like these two projects before. It is no exaggeration to regard the event as the opening of a new epoch in New Zealand's development."(7)

Analysis of proposals

Toward the end of 1954, proposals were received from the New Zealand power consultants and the United Kingdom heavy water consultants. The two proposals were not strictly analogous and there was some difficulty in comparing the costs of the two proposals. A brief summary of the proposals follows.

The power consultants' proposal initially called for a power station of 26 MW which would be capable of expansion to 53 MW. The power consultants proposed that at first there would be two turbine generators of 12 MW, each receiving steam at 40 pounds per square inch gauge (psig) and exhausting to jet condensers at a vacuum of 1.5 inches (3.8 centimeters) of mercury. One turbine generator of 2-3 MW would receive steam at 120 psig and exhaust into the 40 psig manifold, or into the atmosphere. Initially, steam from ten low-pressure wells and ten high-pressure wells were allowed for. The low-pressure wells would discharge at 70 psig well-head pressures, and the high-pressure wells at about 150 psig. Separators would be installed at each well head to separate the steam from the water. In the initial stage, the power consultants anticipated that there would be two steam trunk lines 27 inches (68.6 centimeters) in diameter conveying steam from the bores to the powerhouse. These would operate at 60 psig and 150 psig, respectively, and would also suffice when the plant was expanded to 53 MW. They would provide the steam for the turbine generators operating at 40 psig and 120 psig, respectively, the steam first passing through separators. The initial stage did not call for the utilization of flash steam from separated bore water, which was discarded; but for the 53-MW station, the separated water from high-pressure bores at 150 psig would be at 40 psig and flash steam conducted to the trunk lines served the low-pressure bores.

The powerhouse would be situated fairly close to the river. Consequently, the operating floor and machine foundations would have to be raised about 45 feet (14 meters) above the ground so the barometric legs of jet condensers would not be endangered if river levels rose.

The power consultants' proposal said that cooling water could be pumped from the river through a pipeline and returned to the river through a concrete canal. They said that a part of the cooling water system would serve both stages, but pumps, pipelines, and other equipment would have to be duplicated for the 53-MW stage.

The second proposal, from the heavy water consultants, called for a power station of 37.5 MW plus a heavy water plant to produce 6 tons of heavy water per year, the whole plant being capable of expansion to twice this size. Initially, the power station would contain three identical turbine generators of 12.5 MW each, receiving steam at atmospheric pressure and exhausting to jet condensers at 1.5-inch (4-centimeter) mercury. The proposal envisaged that 18 wells, discharging at a well-head pressure of 70 psig, would serve the intial plant, and separation of steam and water would take place at each well head. Steam would be transported to the plant site through a 60-inch (1.5-meter) pipe and would pass through a final separator before going to the heavy water plant. Separated water would be transported to the plant site through a 30-inch (75-centimeter) pipe, passed through scrubbers and separators, and then be conducted to one of the turbine generator sets. The separated water would be discarded. Both the 60-inch (1.5-meter) steam truck pipe and the 30-inch (76-centimeter) water trunk pipe would serve the needs of the expanded plant. Exhaust steam from the heavy water plant at atmospheric pressure would supply two of the turbine generator sets. Condenser cooling water would be pumped from the river through an 84-inch (2-meter) pipe and returned through an 8-foot (2-meter) concrete line tunnel. Both the pipeline and tunnel would serve the expanded plant, but pumps and other equipment would have to be duplicated.

The powerhouse would be situated further away from the river than in the power consultants' proposal. The plant location had been selected to make the most effective use of natural ground levels for the barometric legs of condensers, which are vertical pipes that drop from the condenser's outlets to a hot well approximately 45 feet (14 meters) below the ground with a tunnel leading from the hot well to the river. The rise in river levels of 10 feet (3 meters) for future hydroelectric developments had been allowed for.

Comparison of costs

The cost of the power consultants' proposals for a 25-MW and a 53-MW station are set out in table 2.2.

Table 2.3 provides a summary cost estimate for a 6-ton-per-annum and a 12-ton-per-annum heavy water plant with a 37.5 MW and a 75 MW power station, respectively. The two proposals are then compared in table 2.4 to show the cost of each power generation proposal on the basis of dollar/kW installed. The unit costs have been itemized and grouped as far as possible in the tables so that a comparison can be readily made.

Table 2.2. Power Only Proposal: Summary of Estimate of Cost (in £)

	26 MW	53 MW
Land	0	0
Transmission line and 220-kW switch gear	224	225
Indoor switch gear and transformers	191	336
Main buildings	230	307
Circulating water works and pumphouses	61	108
Turbo-alternators with condensers	460	920
Steam pipes, valves, lagging, separators	137	146
Cranes	13	13
Drilling of bores, casings and head valves	Not charged	
Workshop equipment and furniture	40	50
Village	130	130
	1,486	2,305
Engineering and contingencies (15 percent)	223	346
	1,709	2,651
Interest during construction (1.5 years at 4 percent)	103	159
	1,812	2,810
Cost per kW installed	£70	£53

Discussion of proposals

The proposals from the two consultants were different in several respects. In order to get a clearer understanding, the Commissioner of Works, the permanent head of the Ministry of Works and Development, suggested that the consultants either reconcile the reports or explain the reasons for the differences. Accordingly, the consultants set up a working party which met to try to come up with an agreement. The fundamental difference between the two reports arose, the working party explained, because of the different approaches of the two groups. The power consultants' report was based on the production of power only, the heavy water consultants' report emphasized heavy water, and whatever power that was generated by its production would be merely a by-product.

The power consultants' report kept very much in mind the uncertain fact of corrosion, erosion, and the technical aspects of steam production. They also addressed the question of the need for electricity generation as a continuing process; electricity generation would grow as the demand grew. Their report therefore emphasized the advantage of

Table 2.3. Allocation of Capital Costs for 6- and 12-Ton-per-Year Heavy Water Plant with Power-Generating Equipment (in £)

Item	6 Tons per Year 37.5 MW		12 Tons per Year 75 MW	
Steam Bores and Distribution				
Bores	450		900	
Offsite steat piping	410		562	
Contractors' charges	152		247	
		1,012		1,909
Cooling water supply				
Offsite piping	150		227	
Contractors' charges	27		38	
		177		265
Heavy water plant				
Materials, freight, erection	732		1,441	
Common items	308		324	
Contractors' charges	185		298	
		1,225		2,063
Generating plant (exhaust steam from heavy water plant)				
Materials, freight, erection	940		1,803	
Common items	560		573	
Tunnel to river	58		58	
Transmission line	200		200	
Contractors' charges	313		447	
		2,071		3,081
Generating plant (flash steam from bore water)				
Materials, freight, erection	390		750	
Offsite hot water piping	190		274	
Contractors' charges	103		173	
		683		1,197
Totals		5,168		8,315

Table 2.4. Itemized Costs (£/kW Installed)

	Power Only		Heavy Water and Power	
	26 MW	53 MW	37.5 MW	75 MW
Turbo alternators with condensers	26.6	21.2		
Cranes	.6	.3		
Indoor switch gear and transformers	8.9	7.7		
Main buildings	10.8	7.1		
Workshop equipment and furniture	1.9	1.2		
	43.8	37.5		
Generating plant using exhaust steam from heavy water plant – materials, freight, erection			29.5	28.2
Generating plant using flash steam from bore water – materials, freight, erection			12.1	11.7
Common items			10.5	5.4
			52.2	45.3
Circulating waterworks and pumphouse	2.9	2.5		
Steam pipes, valves, etc., lagging	6.4	3.4		
	9.3	5.9		
Offsite piping and pumps for cooling water supply			4.7	3.5
Tunnel to river			1.8	0.9
Offsite steam piping			7.7	5.3
Offsite hot water piping			6.0	4.3
			20.2	14.0
Steam bores	nil	nil	8.5	8.4
Village	6.1	4.6	7.1	3.5
Transmission line	10.5	5.2	6.3	3.1
	16.6	9.8	21.9	15.0
Total cost (£/kW installed)	69.7	53.2	94.3	74.3

flexibility to expand the installation to meet future needs. It was the power consultants' view that the heavy water report took too little account of the practical difficulties.

Clearly, the different approaches inherent in the power-oriented as opposed to the heavy water-oriented proposals made it difficult at times to reach a consensus. The appraisal by the UKAEA of the possible future needs for heavy water lead to an announcement in July, 1954 by the New Zealand government of the decision to go ahead, subject to satisfactory economic and technical appraisals. These appraisals were carried out and indicated that, on the basis of the information available at the time, the combined project would be a viable enterprise.

PHASE 2: SELECTION, APPROVAL, AND ACTIVATION

Activation

Geothermal Development Limited

In late 1954, a new organization was created to take over the project. A company called Geothermal Development Limited (GDL) was formed with the object of planning a combined plant to produce heavy water and generate electricity. Two-thirds of the shares of the company were held by the New Zealand government and one-third by the United Kingdom Atomic Energy Authority. The board of directors consisted of three representatives from the New Zealand government and two from the United Kingdom government. The administration of the company was the responsibility of the New Zealand Treasury (see fig. 2.5).

If a joint decision was taken to go ahead with the project, the New Zealand government agreed to purchase all electric power developed at Wairakei; the United Kingdom Atomic Energy Authority, all heavy water. The price for the heavy water was to be based on an amortization period of ten years and was to cover all capital and production costs. At the end of the ten-year period, the heavy water plant was to be passed over to the New Zealand government, which would then become the sole owner of the combined plant.

The MWD found that it could not undertake the civil design work without prejudicing other committed works, and therefore recommended to GDL that a civil engineering consultant be appointed.

With the establishment of GDL, the responsibilities of the various parties, including the different groups of consultants, required clarification. Accordingly, a deed was drawn up between GDL and the New Zealand government in which GDL appointed the heavy water consultants and power consultants to be responsible for design, and both to be assisted by the civil consultant as required. The New Zealand government appointed the MWD to provide the necessary coordination and to supervise construction that had been approved by GDL. It was intended that this arrangement would last until construction was complete, when the combined plant would be operated on behalf of the GDL by the SHED.

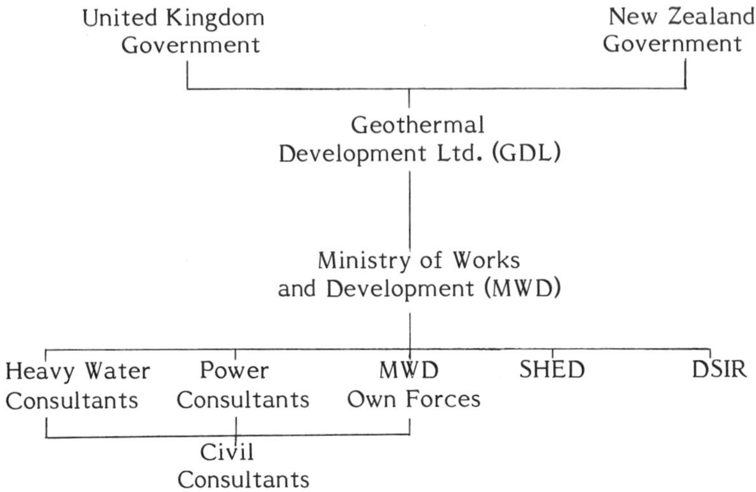

Fig. 2.5. Project organization when the heavy water scheme was included in the project.

To provide the necessary coordination in London, the engineer-in-chief of the MWD appointed a senior official with proven managerial ability to work with the UKAEA and the three groups of consultants.

Revised cost estimates

As the first geothermal power project in New Zealand and only the second to be undertaken in the world, the work at Wairakei was in many respects novel. Consequently, the preliminary estimates, prepared in 1954, were based on design concepts not previously developed in detail. The translation of these concepts into workable form presented problems not commonly encountered, particularly as far as the heavy water portion of the project was concerned.

By the end of 1955, design work had proceeded to the stage where the prices for the equipment, the amount of material needed, and the labor required were fairly accurately known. A revised estimate made at this time doubled the cost of the heavy water section of the project, but increased the cost of the power section by only a third. Table 2.5 affords a comparison of the 1954 estimate and the 1955 estimate, which are essentially a composite of estimates made by the power and heavy water consultants and the GDL.

The final cost of steam could still not be determined. Engineers expected that it would vary with the percentage of productive holes found during the drilling yet to be done for high-pressure steam and with the percentage of investigation costs which GDL could bear. Further, the amount of steam to be supplied had in fact increased by 15 percent, at the request of the Atomic Energy Authority, to give them a better working margin.

Table 2.5. Comparison of Prices for 1954 and 1955
(in £)

	Estimate of November 3, 1954		Estimate of November 18, 1955	
	H. Water	Power	H. Water	Power
Steam service	200,000	571,000	230,000	656,000
Cooling water	–	197,000	–	(incl. below)
Plant	1,607,000	2,330,000	3,555,000	3,654,000
Permanent village	86,000	147,000	120,000	200,000
Interest during construction	108,000	186,000	223,000	258,000
	2,001,000	3,431,000	4,128,000	4,768,000
Percent increase			106	39

Total £8,896,800

Part of the reason for the increase in cost was that the power plant now required had to be 15 percent greater in capacity to make use of the extra steam required by the Atomic Energy Authority. A small increase was due to the escalation of prices since the early estimates had been made. Other costs, such as the cost of the village to house the workers, were underestimated in the 1954 estimates. The single greatest increase in total estimated cost, however, was due to the increase in the cost of the heavy water equipment.

Status of Wairakei defined

The status of the Wairakei project was clearly defined in 1955 by the government through an "order in council" which declared the undertaking to be a "public work." This was considered necessary because, although the government had a real interest in the work, there was no direct appropriation and the project did not fall clearly into any of the many specific definitions of "public work." Government legal advisors considered that even though the generation of electrical energy was certainly a public work and the development of geothermal energy might be argued to be a public work, doubt may have been raised because of the inclusion of the GDL in the arrangements. The order in council was intended to clear up these doubts.

Withdrawal of UKAEA

Toward the end of 1955, the planning of the heavy water plant reached the stage at which reliable estimates of cost could be made. These estimates were much higher than the preliminary ones made by the Atomic Energy Authority and its advisors in the absence of any experience with such a plant. In January 1956, the UKAEA decided that in view of the doubling of the expected capital cost of the project, the unit cost of the heavy water to be produced would be seriously uncompetitive with alternative sources. They decided, therefore, not to continue with their participation in the heavy water project (see fig. 2.6).

This decision came as a blow to the New Zealand partners because Geothermal Development Limited had been set up to accommodate that part of the project, and it was an important part of the overall geothermal development. When the decision was made to abandon the heavy water project, the power consultants were asked to prepare a preliminary report on a project for power only, using equipment already designed as far as possible. Throughout the design phase and the discussions on the overall estimates, the heavy water portion of the project had been a major part of the decision-making process. Adjustments to the original design concept were made, and by April 1956, the main characteristics of the first power station to be built had been already settled and more detailed design work was proceeding.

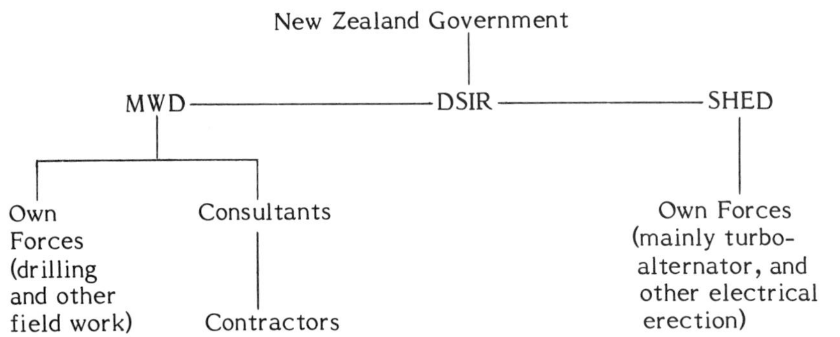

Fig. 2.6. Project organization at the start of power station construction.

The decision was made that the "A" station would house the high-pressure and low-pressure turbo-alternators already being manufactured in Britain. These had a total generating capacity of 46 MW. In addition, two intermediate pressure machines were to be installed, bringing Stage I's development output to 69 MW. The intermediate machines were designed to use the part of the energy in the steam which was formerly allocated to heavy water production. The powerhouse containing the machines would be made large enough to contain two more high-pressure machines to be added as steam became available.

Subsequent extensions

Stage IA involved one 11.2-MW low-pressure machine to run on steam obtained from the large quantities of hot water running to waste from the bores.

It was evident from the results of drilling at the time that this revised proposal was being prepared that there was every likelihood of finding additional steam. Accordingly, provision was made for two additional machines to be installed in "A" station.

Construction of Stage I and Stage IA were not far advanced when the consultants were asked to prepare proposals for additional generating capacity because more steam than anticipated had been found. This resulted in the installation of two more 11.2-MW high-pressure machines in "A" station, and three 30-MW mixed pressure sets in "B" station, which was to be located some 100 feet (30 meters) from "A" station. This brought the total installed capacity to 192 MW.

The proposed extensions allowed for a third stage, which would have consisted of two more 30-MW mixed pressure sets in "B" station, bringing the total installed capacity to 252 MW. This stage was not installed, however, because it became evident that there would be insufficient steam to make use of these machines. (For a description of the major features of stages I, IA, and II, see the discussion of economic evaluation in the section, "Phase 4: Evaluation and Refinement.")

Following agreement on the requirements of SHED, the department which would eventually own and operate the station, the power consultants drew up plans and specifications on which tenders could be called. After further study by both the SHED and the MWD, the consultants called tenders. The consultants evaluated the tenders received from contractors and forwarded them to the secretary of the MWD tenders board. After receiving additional input from the MWD and the SHED, the tenders board then made a recommendation to the minister of Works and Development.

This procedure followed normal MWD practice, which required all tenders to be accepted on behalf of the Crown by the minister of Works and Development.

PHASE 3: OPERATION, CONTROL, AND HANDOVER

Implementation

Status of work (mid-1955)

Until mid-1955, the work at Wairakei had gone strictly in accordance with the construction time schedule set up by the engineer-in-chief of the MWD and the consultants, which called for the production of heavy water at the end of 1957 and power production in June, 1958. The schedule was a difficult one, however, and sticking to it required careful planning.

By the middle of 1955, work at the site included foundation testing for the powerhouse, survey of the pipeline route, and survey of the site generally. The general layout had been planned and the planning of the permanent village was well under way. Construction was expected to start on the village shortly after the middle of the year. The construction site had been tentatively planned to allow for contractors and storage. As for the actual power station activities, drilling of holes for steam production was proceeding and it was expected to obtain the power equivalent of 60,000 kW which was necessary for the first phase of the planned project. The power consultants proposed that there should be some reserve on the planned output and that drilling should continue at an accelerated rate so as to produce sufficient steam to enable the second plant to get underway. The design of the first plant was being arranged to facilitate the extension of the second plant.

The MWD had called for international tenders for the turbines and alternators. Inquiries were made through MWD contacts in Europe and in New Zealand about the steel work for the heavy water towers.

Many resource persons were involved, including the consultants, the British government agencies such as the Atomic Energy Authority, and the New Zealand government departments. The New Zealand engineers in charge of the project recognized that there was a large amount of construction work to be done and a considerable degree of coordination was necessary if the project were to run smoothly.

Site construction

The first contracts, for the turbo-alternators and other associated equipment, were let in mid-1955. Fabrication of this equipment was well advanced when the heavy water side of the project was dropped. Tenders had also been called for the site contracts but had not been let.

After the necessary changes had been made to the design after the UKAEA pulled out, tenders for the site construction were called again. The MWD engineers and consultants recognized that no single New Zealand contractor would be able to cope with all the work, and it would probably have to be done by a consortium. Overseas contractors were interested and would be in a strong position to win the contract. Accordingly, an important condition of tendering was that the successful tender would have to include in its group of principals at least one established New Zealand firm. The contractors finally chosen included British and Swiss firms as well as a New Zealand firm.

To help recover some of the time lost, the MWD let tenders for the supply of various materials, including reinforcing steel, aggregate for the concrete work, and the structural steel for the powerhouse. The Ministry of Works and Development completed the powerhouse excavation and backfill. They also provided the initial village for the contractor, including houses, single men's quarters, amenities, and provided a site officer for the consultants' supervisory staff.

The main contractor for Stage I came on site in August, 1956. Because existing local supplies of electrical power were inadequate, the contractor was required to install a temporary powerhouse in which geothermal steam was used to drive reciprocating engines coupled to electrical generators.

Another novel feature of the construction was the use for the first time in New Zealand of the "vertical down" or "stovepipe" technique of welding. All joints in the steam piping were welded using this technique, and all welding was subjected to radiographic examination. These were the main innovations as far as construction work was concerned.

The actual construction of the geothermal power station posed several problems, which can be broadly divided into two major classes: physical and administrative. The physical difficulties arose primarily because of the peculiar nature of the work site, hot water springs in the powerhouse foundations, and excavation work in thermally altered ground. These difficulties were overcome by the close cooperation between design and construction staffs as, indeed, were most of the difficulties encountered during the course of construction. The late delivery of materials from overseas was another physical difficulty which caused long and costly delays in construction.

Administratively, the major difficulty was that advanced planning of work was nullified in many cases by the absence of essential drawings and information as well as the late delivery of materials. This lack of essential information resulted mainly from the redesign required following the withdrawal of the UKAEA's heavy water portion of the project. In particular, labor requirements became difficult to plan with any accuracy, with the frequent result that a particular job was either under- or overstaffed in certain trades.

In light of experience gained during the operation of the temporary power plant, built by the main contractor, installation of diesel-electric generators may have been more appropriate for construction power. The final result was satisfying, though, particularly as the first geothermal electrical power ever to be fed into the national grid, the main power supply for the country, actually came from this temporary power plant. This alone was a distinction which made some of the difficulties of the job worthwhile.

The first site contract covered the construction of Stages I and IA. Stage II was constructed under a separate contract, using a different group of contractors. The Stage II contract included the construction of "B" station, the second powerhouse, and the installation of three 30-MW mixed pressure sets in it; the installation of two 11.2-MW sets in "A" station, four additional cooling water pumps in the pumphouse; and the additional steam collection and transmission system required for the increased steam needed at the station. Stage II also included a minimum amount of work toward Stage III. This was mainly excavation and backfill for "B" station extensions and provisions of extra cooling water disposal culverts. These were activities which would have been difficult to carry out with "B" station operating if Stage III had gone ahead.

The commissioning of the last machine in Stage II in 1964 saw the end of the construction phase at Wairakei. Subsequently, a substantial amount of construction work was undertaken for better utilization of the energy from the wells.

Supervision and Control

The construction of the Wairakei Power Project was divided among three groups. The Ministry of Works and Development was responsible for drilling the wells and associated work which represented a relatively small part of the total cost. SHED erected the turbo-alternators and other electrical equipment, which likewise represented a relatively small part of the total cost. By far the greatest part of the work was undertaken by contractors. They were responsible for the construction of the steam collection and transmission system, the powerhouse, including machine foundations, and the pumphouse and water cooling system. The work of the contractors was supervised by the consultants, who also exercised some measure of control over the construction of the turbo-alternators. The contract for the supply of the turbo-alternators contained a proviso that the manufacturer was to supply an experienced man to supervise SHED's own forces to erect them. SHED usually provided the construction force, which was composed of the men who would eventually operate the completed station. This procedure was invaluable in making the men thoroughly familiar with the machines they would later operate. In the case of Wairakei's turbo-alternators, the power consultant was still responsible for matters of contract administration, technical problems involving other sections of the work, and the like.

Certain materials had been procured by the MWD, and these were on site before the main contract was awarded. The MWD had also arranged for the excavation of the powerhouse foundations. Subcontracts for the supply of other materials and equipment had also been arranged before the contract was let.

All major subcontracts for material and equipment were let through the MWD tendering system on the basis of drawings and specifications prepared by the consultants. The consultants were responsible for ensuring that the deliveries promised by the subcontractors were on schedule. It was the main contractor's responsibility, however, to ensure that subcontracts for erection on site were on schedule. In effect, he had to include them in his own planning.

Administration of the site contract was wholly the responsibility of the consultants. No directions were issued to the contractor by the MWD. In the event that the MWD had doubts about any aspect of the work, these were discussed with the consultants, who then initiated the required action.

Progress was reviewed monthly at site meetings attended by representatives of MWD, SHED, the consultants, the contractors, and the major subcontractors whose progress was affecting or likely to affect the work at any given time. Each party was required to report briefly on his own progress, emphasizing particularly matters likely to affect the progress. The consultants also reported on the progress of offsite subcontracts. The MWD engineer in charge of the Wairakei project was the relatively unbiased chairman of the meetings. The progress meetings did not relieve the consultants of their responsibility to keep the overall progress continually under review.

In the event a dispute developed between the consultants and the contractor, attempts were made to solve the problem on the site, using a member of the MWD as moderator. If reconciliation was not possible, the engineer-in-chief of the MWD, who was engineer to the contract, and as such was formally in charge, was approached. If this attempt at reconciliation failed, an independent arbitrator was retained – that is, if agreement could be reached on a suitable arbitrator. If all else failed, the contract provided that the Arbitration Act be invoked, which was a lengthy and expensive legal process. Few disputes arose, however, which could not be resolved onsite. The only major items were the inevitable claims for extras – more work or more materials – and these claims were resolved with the assistance of an independent arbitrator.

Completion and Handover

Following the usual practice for the construction of government-owned power projects in New Zealand, the responsibility for construction of Wairakei was shared between MWD and SHED. The exact division of areas of responsibility varies from project to project, but in general, the MWD is responsible for the design and construction of all civil engineering work and a large part of mechanical engineering work, either with its own forces or through consultants and contractors. SHED is responsible for the design and construction of electrical work and the balance of mechanical work. SHED will normally use the consultants employed by the MWD, if any, and work through the MWD tenders board for contract administration.

One aspect in which SHED always becomes involved is the construction of the turbo-alternators and other electrical equipment. They will normally undertake these tasks with their own forces, composed of men who will eventually operate the completed station under the supervision of an experienced erector supplied by the manufacturer. In the case of Wairakei, this arrangement proved invaluable in making the men thoroughly familiar with the machines they were later to operate.

On completion of the project, SHED assumes full responsibility for operation and maintenance. They look after minor maintenance themselves but retain MWD for major surveillance and maintenance work.

Completion date

Wairakei was constructed in two stages, involving some 13 turbo-alternators. Installation of the turbo-alternators was completed progressively, with the steam supply keeping pace. As the various machines were completed, they were operated and their energy fed into the grid.

There was thus no single well-defined completion date on which the works were formally handed over to SHED. Two significant dates, however, are November, 1958, when actual operation commenced with the first power from Wairakei begin fed into the national grid, and October, 1964, when the thirteenth turbo-alternator was commissioned, completing, for the most part, the construction phase.

Rundown of construction activities

Clearly, construction activities at Wairakei and other large projects do not suddenly stop. They run down relatively slowly. This extended rundown means that construction forces can be redeployed to other projects, or can seek other employment. There is no sudden surplus of labor on the market, nor is there a sudden change in the community generally.

Construction plants and equipment can also be redeployed to another project, or they can be sold or scrapped. On a large project, there may not be much economic life left in the plant and equipment at the end of the job. At Wairakei, the contractor's plant was removed from the site with a certain degree of economic life remaining. As far as the MWD plant is concerned, the major items, such as the drilling rigs, are still in use today.

Maintenance

While SHED retains MWD for major surveillance and maintenance work, there was no formal agreement concerning maintenance on the project once it was operational. However, informally, the arrangement was that MWD carried out all maintenance on the wells below the master valve and such other maintenance on an installation of surface equipment as requested by SHED.

MWD monitored the field performance in conjunction with DSIR, and advised SHED accordingly. Proposals for maintenance work on wells were generally agreed upon with SHED before the work started, although in an emergency, MWD acted first and obtained approval afterward. This procedure still holds true today at Wairakei.

There is no fixed surveillance program for surface installations other than to ensure that all producing wells are visited at least once a day, and all others at reasonably frequent intervals.

PHASE 4: EVALUATION AND REFINEMENT

Technical Evaluation

The two major uncertainties with Wairakei at the conceptual stage were the probable life of the field and the effects of corrosion. The decision made at the outset to go ahead in spite of these uncertainties was later justified by the consistent and reliable performance of the station.

Station output

Initial power from Wairakei was fed into the North Island grid in November 1958, but due to development of the project in stages, commissioning was not completed until October 1964. Table 2.6 sets out the annual energy generation from Wairakei, together with the contributions in percentages that this had made to the SHED generation in the North Island and New Zealand as a whole.

Apart from the substantial contribution Wairakei has made to New Zealand's energy requirements, the main feature table 2.6 reveals is the consistent output from the station. Except for 1968, a special case, the annual generation has been maintained at 1,200 $\pm$50 gigawatts per hour (gwh) since commissioning was completed. This has been accomplished by making more efficient use of the discharge or, in other words, by improving the utilization factor.

The utilization factor, a measure of the overall efficiency of the station, is defined as the ratio of electrical energy generated, expressed in heat units, to the total heat energy above 32°F (0°C) discharged from the field. It reflects not only the usual thermal and mechanical efficiencies, but also the considerable amount of energy contained in the waste water.

The last turbine in the power station was installed in 1963, bringing the installed capacity to 192 MW. Originally, the plan was to use one of the 30-MW machines and one 11.2-MW machine as spares, but later it was hoped to supplement the intermediate-pressure (IP) and low-pressure (LP) steam supplied from hot water transported to the station and there flashed to steam. A plant for this purpose was commissioned in 1963.

The hot water production from the group of wells supplying the flash plant declined rapidly, however, and the plant was closed down in 1964. A program of well drilling to maintain steam supply ran from 1964 to 1966. Diminishing returns from this program led to the realization that it would be more economical in the long run to limit field production, and no new production wells have been drilled since.

Instead of drilling, several strategies have been adopted. High-pressure (HP) well-head pressure has been progressively reduced from about 200 psi to 135 psi. This resulted in a greater steam production from HP wells. HP turbine output declined because of the reduced turbine inlet pressure, but the extra steam passed to IP and LP sets resulted in an overall increase in station output.

This process maintained a reasonable return of the investment in steam wells by optimizing the power production from HP wells within limitations imposed by station equipment. HP wells were converted to IP ones as this became necessary for individual wells. Waste water from HP wells was increasingly utilized to produce IP steam by the installation of extra flashing equipment in the field. HP steam lines were converted to IP to match steam quantities.

In the latest program, waste water from IP wells and IP flash plants has again been flashed to produce LP steam for transmission to the

Table 2.6. Electricity Generation at Wairakei

Year Ending	Generation in gWh (a)	Percent of North Island by SHED	Percent of New Zealand by SHED
3/31/59	6	0.002	0.001
3/31/60	169	3.9	2.8
3/31/61	384	8.2	5.9
3/31/62	491	9.7	6.9
3/31/63	761	14.0	10.0
3/31/64	1004	16.1	11.6
3/31/65(b)	1194	17.5	12.7
3/31/66(c)	1255	19.7	12.3
3/31/67	1268	18.0	11.5
3/31/68	1058	17.1	9.4
3/31/69	1206	18.0	10.2
3/31/70	1243	18.7	9.9
3/31/71	1185	17.0	8.9
3/31/72	1174	16.1	7.9
3/31/73	1175	15.7	7.0
3/31/74	1162	13.9	6.6
3/31/75	1249	13.2	7.0
3/31/76	1272	13.4	6.5
3/31/77(d)	1232	10.1	6.0
3/31/78	1158	9.4	5.6
3/31/79(e)	1190	-	-

(a) A gWh is equal to 10^6 kWh.
(b) The last machine at Wairakei commissioned in October 1964.
(c) This was the first full year of operation of the interisland transmission system.
(d) The decrease in Wairakei's share of the North Island, and total New Zealand, generation in 1977 resulted from the commissioning of two new thermal stations in the North Island.
(e) Final generation figures for 1979 were not available at the time of writing.

Source: New Zealand State Hydro-Electric Department (SHED), _Annual Reports_, 1959-1979.

station. A new steam line has been constructed for this purpose. Periodically, steam transmission pipework has been changed and new pipes built to match the transmission capabilities with the steam production.

With each increase in the amount of flashing equipment installed to produce steam, more energy has become available from more or less the same total water production and the utilization factor has improved, as shown in table 2.7.

No further gain can be made in the utilization factor, as all the heat being wasted before is now being used. As will be discussed later, the decline in underground pressures and temperatures is substantially less than it was earlier, and the field appears to be approaching a new level of stability. In case it should be necessary, however, the following methods of maintaining station output are now being studied: increasing production from existing wells, producing from greater depths, and maintaining underground pressures.

Table 2.7. Utilization Factor, 1964-1974

Year	Utilization Factor (Percent)
1964	4.66
1966	6.05
1968	6.82
1970	7.23
1973	8.35
1974	8.89

Field performance

The production history of the field extends over nearly 26 years, and in that time, substantial changes have taken place in the underground conditions. Initially, the acquifer was filled with water, with the temperature and pressure conditions following the boiling point for depth relationship until a temperature of $500^{o}F$ ($260^{o}C$) was reached.

Development of this natural resource has resulted in an almost uniform pressure decline of over 21 bars, affecting an area considerably greater than that area in which the production wells were drilled. This area is defined approximately by the Wairakei Stream in the north and northeast, the Waikato River to Huka Falls in the southeast, the Waipouwerawera Stream in the south, and Poihipi Road in the west. These features may be seen in figure 2.7.

Figure 2.8 shows the relationship between the rate of pressure decline and rate of drawoff to the end of 1968. As can be seen, this is not a linear relationship in that while there is an immediate fall in

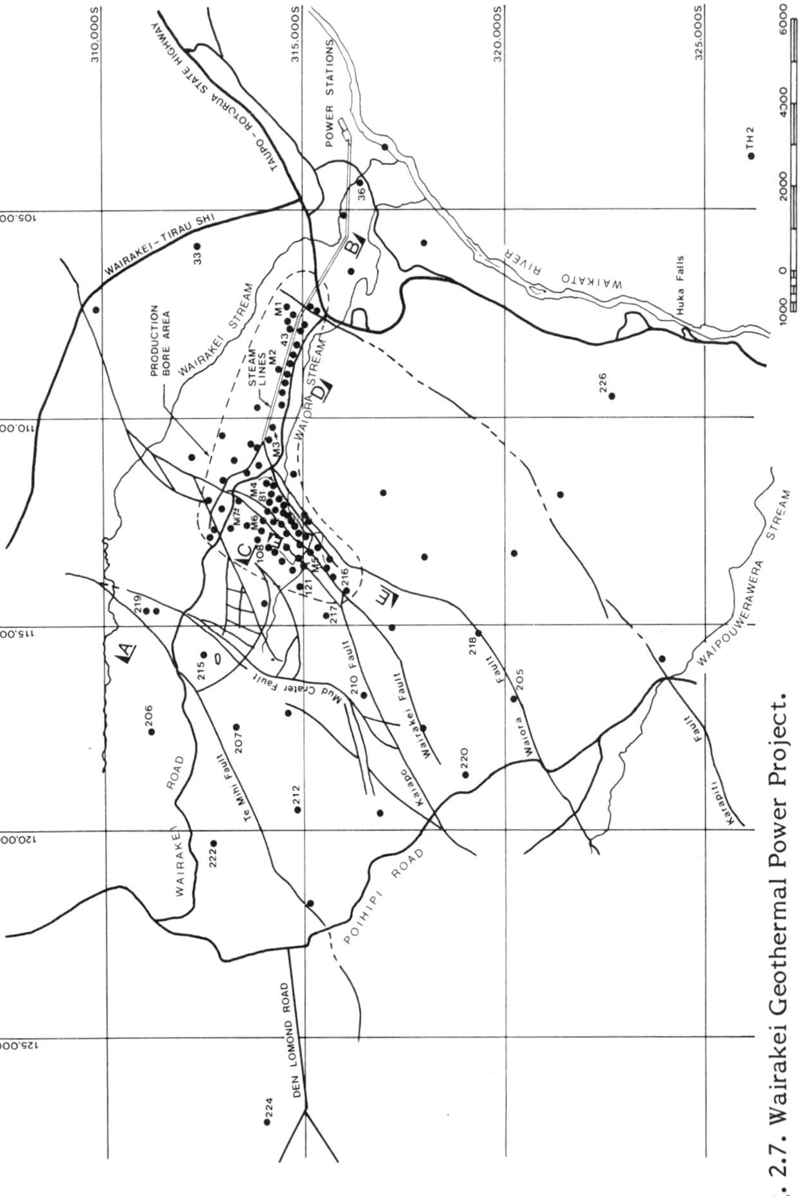

Fig. 2.7. Wairakei Geothermal Power Project.

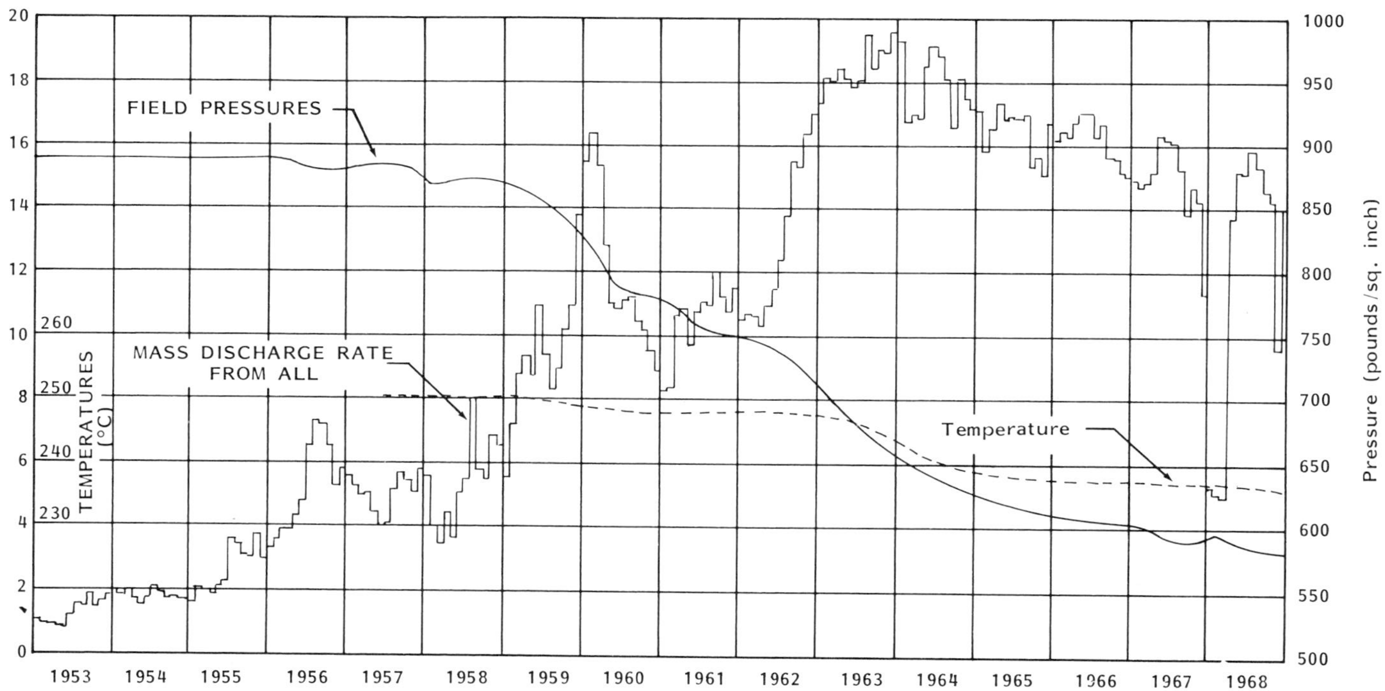

Fig. 2.8. Field discharge rate, pressures, temperatures, Wairakei. (Pressures shown are those at a level 900 feet below sea level, Reduced Level or R.L - 900)

pressure following an increase in drawoff, a prolonged period of substantially constant drawoff shows pressures tending toward a stable value. The relation between pressures and drawoff suggests that conditions are being influenced by an inflow of geothermal water, which is supported by the rise in pressure at the beginning of 1968 in a period when the field drawoff was reduced to about one-third of normal. Since 1968, pressures have fallen by a further 30 psi.

Temperature trends in the upper levels show a similar pattern as is also illustrated in figure 2.8. The temperature trends shown are for the average of the maximum temperatures in the wells in the western or main production area, and reflect the temperature in the production area. Because it implies that the temperature of the inflow has not changed, it is important to note that temperatures at greater depths have shown no change.

Trends in individual well outputs are shown in figure 2.9. This shows clearly the effect of the fall in pressures and temperatures in the producing zone. Extrapolation of the trends in total discharge from the field, however, indicates that the flow tends toward a stable value sufficient to maintain an output of between 125 and 140 MW indefinitely. This compares with a present (1979) output of 148 MW.

Some wells show an increase in steam content, while others show a decrease. Indeed, the enthalpy of the total discharge is, if anything, tending to decrease. Thus, while the possibility of the field going dry cannot be discounted, it would obviously be a long-term process of which there is no indication at the present time.

For Wairakei, forced outages – those that are the result of unforeseen failures – were almost negligible, and the station was in service 85 percent of the time, generating 80 percent of the energy which could be generated were it run at its maximum continuous rating. These figures were the highest for any station in the country, hydro or thermal, and were significantly higher than those for other thermal stations. Wairakei has maintained this record since commissioning was completed in 1964.

Corrosion

As previously noted, chemical impurities in the geothermal fluids required some special care in the selection of materials which come into contact with them. An extensive series of field tests on material samples initiated in 1950 yielded valuable information concerning this problem. Mild steel has been proven to show good resistance to geothermal fluids as long as no oxygen is present. This material can therefore be safely used for well-head gear, steam and hot water pipes, flash vessels, and other equipment. The initial Wairakei design called for the addition of 0.12 inch (0.30 centimeter) extra wall thickness as a corrosion allowance. As a result of experience, however, this requirement has been found to be unnecessary and was omitted from later modifications to the pipework.

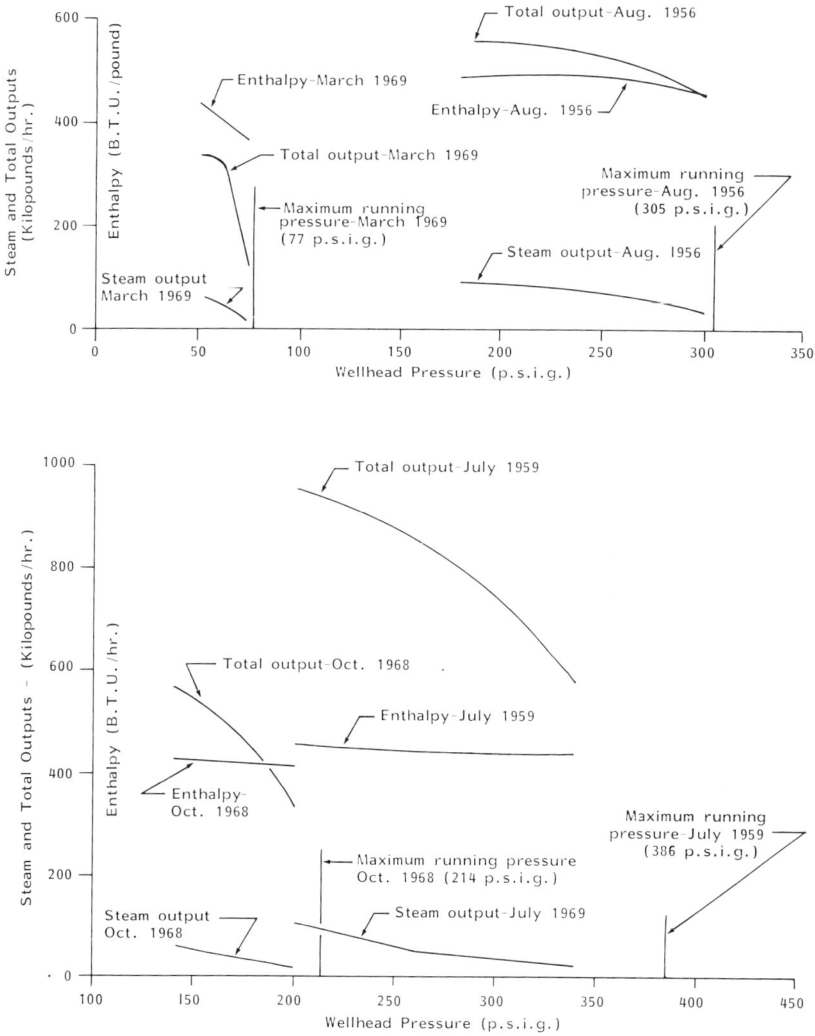

Fig. 2.9. Wairakei well output characteristics.

Copper and copper-based alloys (other than certain brasses) are vulnerable to corrosion and thus preclude the use of brazing spelter. It also requires care in the selection of materials for valve parts that contact the geothermal steam or water. Hydrogen sulfide attack on the copper commutators on the alternators requires periodic maintenance and has also resulted in the replacement of copper relay contacts with gold-plated or platinum contacts.

Standby corrosion on stainless steel compensators caused difficulty on one occasion, and stress corrosion cracking occurred on two well-head compensators.

A major influence on the turbine design was the choice of soft stainless iron for the blade material. This type of iron is less susceptible to stress corrosion cracking than harder material, but is more susceptible to erosion. For this reason, blade tip speeds were held at 900 feet/second (274 meters/second). The choice was sound, as there have been no failures resulting from the material used. In fact, it is possible that the blade tip speed could be increased, reducing machine sizes.

The most extensive corrosion was in the condensers and gas exhausting equipment, when the combination of gas, oxygen, and moisture resulted in an extremely corrosive atmosphere. This required use of specially formulated paints, of aluminum spraying, and of stainless steel. In the case of low pressure exhaust gas pipework, engineers found it necessary to use a glass-fiber-reinforced plastic pipe.

The various points mentioned are relatively minor, however. A notable feature of the operation of Wairakei has been the small amount of corrosion which has occurred.

Economic Evaluation

Capital cost

The project was constructed in two main stages, Stages I and II, with intermediate Stage IA. Separate powerhouses were built during Stages I and II ("A" and "B" stations). The main features of Stages I and II and the years during which construction and commissioning were carried out are shown along with details of capital expenditure in figure 2.10.

Stage IA

This stage was a pilot project to investigate the possibility of transmitting hot water from 5 HP and 2 IP wells – water which was previously discharged to waste – and flashing IP and LP steam in a plant alongside "A" station. To prevent boiling in the pipeline, the project involved increasing the pressure by pumping and cooling some of the HP water, already at boiling point, by attemperating with IP water. The pressure was maintained by controlling the water level in an elevated tank at the upper end of the hot water main. The main features were:

Stage I (1956-1960)

Installed plant – 69 MW.
"A" powerhouse.
Turbo-alternators – Two 6.5 MW sets (HP).
Two 11.2 MW sets (IP).
Two 11.2 MW sets (LP).
Cooling water pumphouse with four pumps.
41 wells, of which 28 were connected to the steam transmission system.
Steam transmission system comprising 5- to 20-inch (51-centimeter) diameter pipelines, 3 HP and 2 IP, total length 42,500 feet (13 kilometers).
Hot water drains, roading, and other ancillary works.

Stage II (1960-1963)

Installed plant – 112.4 MW
"B" powerhouse.
Turbo-alternators – Three 30-MW sets, IP and LP inlets, in "B" station
Two 11.2-MW sets, HP in "A" station.
Cooling water pumps – 4 additional in existing pumphouse.
53 wells of which 34 were connected to the steam transmission system and 7 were monitor wells for steam field surveillance.
Steam transmission system comprising extensions to 20-inch diameter (50-centimeter) mains, 2 HP and 1 HP, total length 27,650 feet (8,295 meters).
Extensions for hot water drains, roads, and other steam field facilities

(The total of 94 wells shown for Stages I and II includes all exploration wells.)

Fig. 2.10. Major features.

1. Installation of pumps, vessels, and controls at seven wellheads,
2. Hot water pipeline, 17- to 19-inch diameter (43 to 48 centimeters), totaling 5,300 feet (1,590 meters) in length,
3. Elevated tank and controls,
4. Flash plant vessels, scrubbers, and controls, and
5. Turbo-alternator: One 11.2-MW set, LP in "A" station.

By the time the plant was commissioned, the water output from three of the wells had decreased considerably and only four came into use, increasing the power output by 5 MW. Its feasibility was proved, but the project was later abandoned due to further decline in output from wells and the pipeline used for transmission of IP steam.

Overall planning provided for a possible further development of 90 MW (Stage III), by using steam flashed from the hot water. This was never installed, although some expenditure was incurred in anticipation of installation. Table 2.8 gives the major division of costs, and cost per kilowatt installed, while the capital costs are itemized in table 2.9.

Table 2.8. Major Division of Costs (in $NZ)

Capital	
Exploration	1,300
Exploitation	17,176
Utilization	24,891
Total to 3/31/69	43,367

Note: Cost per kW installed at 3/31/69, $NZ225.37. Cost per kilowatt installed is based on the total installed capacity of 192.6 MW. In fact, because of steam supply limitations, the capacity of the station is reduced to figures which range from 170 MW to 150 MW.

Operating and maintenance costs

Operating and maintenance costs consisted of working expenses, capital charges, and administration costs. Working expenses averaged over a four-year period, 1965-1969, are shown in table 2.10. In addition to normal maintenance, works of a capital nature undertaken to restore or maintain steam output were charged to operating expenses in the year in which they were met and were reflected in the cost of generation. Such work included:

1. Eleven new wells drilled during 1964-1966 at a cost of $940,158,
2. One exploration well drilled to a depth of 7,395 feet (2,254 meters) in 1968 at a cost of $500,118,
3. Steam pipeline modifications,
4. Installation of equipment for flashing IP steam from HP water at some well heads and feeding it into the steam transmission system, and
5. Installation of a 42-inch (107-centimeter) diameter pipeline for IP steam, 3,200 feet (960 meters) long, to augment capacity of the transmission system at a cost of $316,486. (This cost has not been included in table 2.10 because it had not come to charge by March 31, 1969, but the other items are included.)

Table 2.10 includes charges of approximately $40,000 per year by the Department of Scientific and Industrial Research for scientific work in the fields of chemistry, metallurgy, geophysics, and geology in connection with Wairakei.

Table 2.9. Capital Costs (in $NZ)

	Stage I	Stage II	Total
Land acquisition	nil	nil	nil
Site preparation	178	100	278
Establishments costs: camps, temporary housing, workshops and offices, and site leveling	662	1,026	1,688
Power stations			
Buildings, foundations, cooling water culverts, offices and workshops, roads, and landscaping	2,982	1,692	4,674
Turbines and generators	3,413	5,461	8,874
Steam pipework and valves	508	459	967
Electrical work: 400-volt and 11-kilovolt switch gear, panels, control room, transformers other than generator transformers	480	360	840
Generator transformer	298	278	576
Outdoor structure and switch gear, high tension circuit breakers	398	266	662
General services electrical, air, telephones, etc.	49	66	115
Cooling water pumphouse	1,419	-	1,881
Pumps, mains, valves		462	
Workshop tools	16	2	18
Steam field			
Well drilling (including exploration and unproductive wells)	2,512	5,131	7,643
Well-head equipment	744	894	1,638
Branch lines	387	584	971
Main pipelines	2,970	2,288	5,258
Roads	153	309	462
Main drainage	-	1,641	1,641
Water supply	168	278	446
Landscaping	-	405	405
General services	-	17	17
Permanent village pilot hot water scheme, Stage IA			
Modifications to well-head equipment	213		213
Hot water pipeline	732		732
Flash plant and controls	925		925
No. 10 turbo-alternator	591		591
Work for Stage III (not built)		293	293
Miscellaneous charges and adjustments		241	241
Totals	20,190	23,177	43,367

Table 2.10. Annual Working Expenses for 1965-1969
(in Thousands of $NZ)

Steam Field	Cost
Mineral rights	nil
Land and roads	9.4
Services	6.9
Drains (hot water)	16.8
Wells	
Servicing and modification	84.8
New wells	295.4
Measurements	60.3
Main pipelines	
Servicing and modification	53.7
New lines	30.9
Branch lines	15.0
Steam traps	7.6
Buildings	0.5
Hot water equipment	1.0
Mechanical equipment	2.1
Workshop general	2.6
Operating salaries	25.5
Supervision	8.2
Miscellaneous	1.0
Total for steam field	621.7
Power Station	
Lands and roads	9.8
Services	3.8
Buildings (power station)	19.1
Information center	22.6
MWD buildings	6.2(3 years)
Cooling water system	65.0(3 years)
Gas exhaust system	12.6
Steam lines	7.1
Turbo-alternators	36.5
Electrical equipment	108.0
Other mechanical equipment	25.0
Operating costs	22.0
Supervision	194.6
Miscellaneous	12.6
	43.6
Total for power station	588.5
Total	NZ$1,210.2

The SHED staff of Wairakei was established as follows:

Administration	8
Operation	
Power station	43
Steam field	8
Maintenance	
Power station	50
Steam field	40
Village and services	12
Total	161

The actual numbers of the staff on the site varied from year to year, depending on the workload; the average for the four years was 150 men.

In addition, the Ministry of Works and Development maintains a force of 160 men primarily for investigation work, but who are also available for construction, modification, and maintenance work at Wairakei. Major modification work might be let on contract. These costs are included in table 2.10.

Capital charges were made up of interest, depreciation, and loan repayment, as follows:

Interest on loan liability	5.20 percent
Depreciation	4.59 percent
Loan repayment	0.96 percent
Total capital charges	10.75 percent

Interest rates vary with the source of capital, and 5.20 percent was a mean figure for the year ending March 31, 1969, on the whole of the SHED's loan liability.

Depreciation is based on a formula which assumes a 20-year life span for geothermal stations and provides an increasing amount of depreciation as the asset ages. The effective rate of depreciation for Wairakei for the year ending March 31, 1969, was 4.59 percent.

Administration costs represent Wairakei's share of regional and national administration costs of SHED, based on a formula which takes into account working expenses and staff salaries at Wairakei and the proportion that these items bear to total SHED's working expenses and staff salaries.

Therefore, the total operating and maintenance costs, and the cost per unit of net energy generation, were:

	Cost
Total working expenses	NZ$1,210.2
Capital charges	4,239.2
Administration costs	328.5
Total operating charges	5,777.9
Net generation gHh	1,196.8
Cost per kWh	0.4827 cent NZ

Generation of electrical power in the years ending March 31, 1968, and March 31, 1969 was affected by a partial shutdown of the field during periods of high water inflow into the SHED hydroelectric system, which resulted in a surplus of generating capacity. The shutdowns provided an opportunity for more detailed study of the characteristics of the Wairakei field. The pressure recovery test on the steam field reduced generation to 75 MW for a period of four months. This affected the total generation and cost per unit in this summary.

Environmental Evaluation

When investigations into the possible use of geothermal energy began in 1951, the region surrounding Wairakei was little more than a scrub- and pine-covered wasteland. At that time, the population of the region was about 2,500 people. They lived as isolated settlers, in timber settlements, in Maori settlements on the shores of Lake Taupo, or in the borough of Taupo. Taupo attracted visitors primarily for trout fishing and scenery and for the thermal activity when the project began. The main highway through Taupo had not long been completed when the project began and Taupo had direct access to the north, south, and east of the North Island.

Since 1951, the region has undergone substantial development. In addition to the geothermal development, large areas of land have been developed for pastoral farming, forestry and timber industry have been expanded, road and air access improved, hydroelectric power projects constructed, and at the present time (1979) there is intensive tourist development. All of these activities, together with the supporting services they require, have enhanced the economy of the region considerably. As a consequence, the region surrounding Taupo supports a permanent population of about 13,500.

The Wairakei geothermal field, one of several in the thermal belt on the North Island, covers an area of about 7 square miles (18 square kilometers). The area occupied by the power project is, however, only about 2 square miles (5 square kilometers), not all of which is in the field itself. The following passages describe the effects which the development has had on both the natural and human environments. With one exception, the effects on the natural environment have had impact only in the steam field and in the area developed for power generation. As far as the human environment is concerned, the effects have been more widespread, having influenced considerably the development and economy of the whole Taupo region.

Because the resource being developed at Wairakei is basically the heat from the earth's core, on a human time scale it is theoretically inexhaustible. Since the heat is extracted and brought to the surface by water, however, practical limitations are imposed by the rate at which the water withdrawn can be replenished. The mass and heat outputs from the Wairakei field have declined, but the present rate of decline is less than that in the earlier years of development. In other words, the output appears to be approaching a steady state at which it can be expected to continue more or less indefinitely.

Earth moving and well drilling have been the construction activities most affecting the natural environment. The surface vegetation removed consisted mainly of scrub and exotic pine trees, although some of the later wells and the permanent operator's village were located on farmland, resulting in some loss of grassland.

To our knowledge, the only animals affected by the removal of the scrub and pine were wild pigs. Effects on other forms of life such as birds and insects are unknown, but none has been brought to our attention. Soils, which are of poor quality, were stripped and reused for landscaping. Pipe routes and access roads resulted in some modifications of contours, but had little effect on drainage patterns. Removal of the vegetation led to an increased runoff of surface water, and since the pumiceous material exposed was light and highly susceptible to erosion, this was a problem requiring attention during the construction period. Surface drainage was improved by landscaping and by providing stable channels to control water runoff.

In the early days of the project, it was the practice to discharge the wells vertically for a period to clear them of drilling mud, cuttings, and loose formation material. They were then discharged for long periods through a horizontal silencer. The large quantities of water discharged resulted in considerable local erosion, and any vegetation in the way of the discharge was killed by the salts in the water. Additionally, there was a lot of noise, particularly in the high-frequency range. Present practice is still to discharge the wells vertically for a short period, but discharges are diverted to a silencer of a type developed at Wairakei which gives complete control over the water, no longer causes erosion, and reduces the noise to acceptable levels. In operation, the waste water from 61 wells discharges through the silencers, and substantial quantities of steam are released at atmospheric pressure. This does not appear to have had any noticeable effect on the local climate or temperature. Wells discharging waste water produce some carbon dioxide and hydrogen sulfide, but the amount of gas in the atmosphere in the steam field is very small, certainly very much less than is evident in the general atmosphere of the city of Rotorua.

At Wairakei, the separated water and condenser effluent are both discharged into the Waikato River. Under mean flow conditions, this results in a temperature rise of about 2°C (3.6^{o}F), but it will be greater under low flow conditions.

A problem occurring elsewhere in the Waikato River is a general acceleration of aquatic weed growth. The problem is being studied;

factors suspected of influencing the weed growth include the increase in water temperature due to the discharge of geothermal waste water, enrichment of the river water by fertilizers and other minerals from development, the ponding of hydro lakes and consequent temperature rises.

The effect of the waste heat on fish life is minimal because the hot water is localized and occupies only a small part of the waterway before being dissipated.

In New Zealand, fresh water salinities higher than one tenth that of sea water are unusual. Chlorides discharged from the Wairakei field into the Waikato River do not appear to present any problem. More potentially harmful constituents include silica, arsenic, mercury, and boron in varying proportions. Chemical analyses of the effluents from the Wairakei and Broadlands fields appear in table 2.11.

Table 2.11. Chemical Analyses of Effluents from Wairakei and Broodlands Fields

	Milligrams per Kilograms
Lithium	13.2
Sodium	1,200
Potassium	198
Rubidium	3.0
Caesium	2.7
Calcium	17.5
Magnesium	0.005
Fluorine	8.0
Chlorine	2,150
Bromine	5.7
Iodine	0.5
Sulfate	25.2
Arsenic	4.1
Ammonia	0.15
Boron	27.7
Silica	650
Bicarbonate	-
Total flow – liters/second	1,230
Carbon dioxide – tons/hour	1.7
Hydrogen sulfide – tons/hour	0.07

The concentration of chemicals in the waste water is rapidly diluted by the river flow, but nevertheless the salinity of the river water is increased by salts in the waste water discharged from the wells. There has always been a substantial natural flow of thermal water into the river over its length from the outlet at Lake Taupo through the thermal region. The chemical composition of this water is similar to the waste water from the geothermal wells, so that the river is naturally contaminated to some degree. As far as the wastes from the field are concerned, the most undesirable contaminant in the river is arsenic, which can reach 0.075 ppm. This is greater than the limit of 0.05 parts/million recommended by the World Health Organization for drinking water. The concentration of arsenic in that part of the river within the bounds of the thermal region fluctuates, however. Increases occur according to thermal inflows, and decreases occur due to absorption of the arsenic by sediments, which settle out in the various hydro lakes and concentrate in the water weed. By the time the river flows from the thermal region, the arsenic content is below 0.02 parts/million.

Experience shows that geothermal development does in time reduce or eliminate some natural geothermal phenomena. The disappearance of the Geyser Valley hot springs at Wairakei is an example. The effect on other phenomena, however, may be slow to appear or may be insignificant. Much of the steaming ground around Wairakei still remains active. Moreover, there is a history of natural changes in geothermal features both in New Zealand and overseas (such as Great Geyser, Iceland), showing that it is not reasonable to attribute all such changes to the effects of the development.

Continued operation of the wells has resulted in changes in underground pressures and temperatures, as a result of which subsidence has occurred. The subsidence at Wairakei has a strong vertical and horizontal component and has formed a roughly elliptical dish-shaped depression. The maximum rate of subsidence has been about 1.31 feet (39.3 centimeters) a year, with a total maximum subsidence estimated to be over 9.84 feet (2.9 meters). The center of the subsidence is about 1,476 feet (443 meters) from the nearest wells, and 5,906 feet (1,800 meters) from the region of greatest drawoff. The movement is attributed to the bending of the mudstone cap rock resulting from the fall in pressure and the withdrawal of the large mass of fluid. The geology of the area of maximum subsidence is not precisely known, but there is a direct correlation between the subsidence and the thickness of the breccia which underlies the mudstone in the region for which geological information is available. Precise measurement taken by tiltmeter during a partial shutdown in 1968 suggest that the subsidence may be reversible.

The ground movement has affected steam mains and drainage channels, causing some inconvenience, but without jeopardizing the operation of the field. The effect of the movements measured at Wairakei on powerhouses or similar structures, however, would be little short of disastrous. For this reason, the possibility of ground movement must always be considered, and a comprehensive system of bench marks should be installed in sufficient time to enable potential areas of subsidence to be located before a powerhouse site is selected.

The main effects of station operation on the natural environment result from the discharge of waste gases and the discharge of the condensate and cooling water. The steam separated from the water at the well heads contains all of the gases. These pass through the turbines with the steam, and are extracted from the condenser and vented to the atmosphere. Over the years of development, the amount of gas in the discharge has steadily declined; at the present time (1979) it is less than 67 pounds (30 kilograms) per minute or 0.18 percent by weight of the steam. The gas composition is approximately 95 percent carbon dioxide and 5 percent hydrogen sulfide, with traces of methane, ammonia, and hydrogen. Hydrogen sulfide is evident in the atmosphere in the immediate vicinity of the power plant on those days when there is little or no wind. Concentrations are well below toxicity limits, but the gas increases the maintenance required on electrical contacts. Since both carbon dioxide and hydrogen sulfide are heavier than air, it has been necessary to ensure that adequate ventilation is provided in all areas where leaking gas is likely to accumulate.

The combined cooling water and condensate flow total approximately 350 cubic feet/second (cusecs) (10.5 cubic meters), and it discharges into the river at 91°F (33°C). The net increase in river temperature is approximately 1°C (1.8°F). The total increase due to the combined effects of the waste water, the cooling water, and the condensate is 1.5°C (2.7°F) assuming no loss to the atmosphere. This is less than the annual variation of the temperature of the river, which is about 9°C (16°F). Apart from being slightly acidic due to the small amount of gas retained in solution, the cooling water and condensate discharge add no chemicals to the river.

Two blowouts, or uncontrolled discharges, have occurred at Wairakei. One was due to failure of a well casing after some years in service, the other occurred during the course of drilling a well. Unlike similar occurrences in oil fields, the main effects are restricted to the immediate vicinity of the blowout, although secondary effects may be evident over a wider area.

Environmentally, the main effects of a blowout are the waste of a resource, increased discharge of heat and water locally, and possible danger to other structures in the vicinity. One of the blowouts at Wairakei was remedied by pumping cold water through a deviated well drilled to intersect the well causing the blowout. The uncontrolled discharge was thereby stopped and the offending well filled with cement grout. The other was not stopped, because it was isolated, with nothing in the vicinity likely to be endangered. There was thus a continued loss of energy, and a discharge of steam to the atmosphere. A secondary effect evident during this blowout resulted from vigorous boiling which occurred underground. This created ground vibrations which were quite strong in the immediate vicinity and which made accurate surveying impossible for distances up to one kilometer (0.62 mile) away. Indeed, the blowout was a spectacular tourist attraction until it ceased discharging.

The development of the Wairakei project has converted a small area of wasteland into a productive area, having, as a secondary result, a considerable value as a tourist attraction. Some 100,000 people a year visited the information center in the steam field in 1978 and there are many more who view the project as they pass by. From a scenic point of view, there has been some reduction of natural geothermal activity. The absence of the chimneys, boilers, and fuel-handling equipment associated with conventional thermal power stations has permitted clean and compact designs for the Wairakei powerhouses, which are relatively unobtrusive.

Land use in the area has been affected to the extent that had the geothermal project not been undertaken, the land would probably have been developed for pastoral farming. The productive value of the land as a geothermal project is, however, much higher than it would be as farmland. The surrounding area is currently being developed as a tourist park, and the presence of the geothermal development places some restrictions on this development, particularly as some areas are closed to the public by MWD activities.

The discharge from unsilenced wells is very noisy, and ear protection is necessary in the immediate vicinity. The silencers used are efficient, however, and while there is still some noise, this has been reduced to acceptable levels. The water from 61 wells is being continually discharged to waste through such silencers, but in spite of this, the important Wairakei Tourist Hotel, located across the road from the field, never lacks patronage. Noise levels within the powerhouses are fairly high due to the operation of the gas-exhausting equipment, and the operating staff finds it more comfortable to wear ear protectors while on shift. The noise level is not high enough to compel the use of ear protectors for short periods of exposure, although they are available for visitors if required.

State Highway 1 bridges the steam mains, and in general there are no traffic restrictions on this road. Should restrictions be required, alternative routes are available in the project area. In a few cases, local farmers have had to accept new and longer access roads, but some compensation was provided by improved alignment and surface of the roads. Access to the field is subject to certain restrictions, imposed with the welfare of sightseers in mind, but this inconvenience is minimized to some extent by the availability of guided tours.

Drifting steam occasionally presents a hazard on the roads within the steam field and has led to accidents. The occurrence of drifting steam depends to a large extent on weather conditions. In the early days of the project, a similar problem was encountered on State Highway 1, but this was eliminated when the wells adjacent to the highway were connected to the powerhouse and when their silencers were extended. Warning signs were also placed alongside the highway.

The project has had a strong influence on the regional economy and community patterns. Construction of the powerhouses was carried out in stages, resulting in a lengthy construction period. The peak labor force directly employed on the work was about 600. This meant that a substantial amount of money came into the region. A large part of this was spent in Taupo, providing a stimulus for the development of shopping facilities and other social and cultural needs. A permanent work force of about 225 is directly associated with the power project.

About 150 of these permanent employees are housed in a village built at Wairakei for the power station staff. This has meant the establishment of a school, community hall, and limited shopping facilities. The village serves as a social center for the area in the immediate vicinity. The remainder of the employees are housed in Taupo, which is the main center for the region. There is little doubt that the land occupied by the geothermal project supports a far greater population than it would as farmland, tourist park, or forest.

As far as the national economy is concerned, Wairakei produces 13 percent of the electrical energy consumed on the North Island. In the sixteen years since commissioning was completed, the annual production has been 1,150 gWh, and the station operates at a load factor of 88 to 89 percent. The energy costs less than 0.8 cent/kWh. Other than expenditure for servicing loan moneys, there is no annual outgoing of foreign exchange.

Expertise gained in the development of Wairakei has always been made available to anybody interested without restriction. Many overseas engineers and scientists have visited the project to learn the techniques developed at Wairakei. And New Zealand engineers and scientists have traveled extensively to advise and assist in geothermal developments in other countries. There has been a small economic gain to the country from such consultation and with the rapidly increasing interest in geothermal energy overseas, this is expected to increase. A more intangible gain, but one of considerable value, has been the enhancement of New Zealand's prestige overseas through the assistance which has been given.

This review of the effect of the development of the Wairakei Geothermal Power Project on the natural and human environment shows that there have been both adverse and beneficial effects. The most harmful effects are believed to be the discharge of waste heat and water into the Waikato River and waste gases into the atmosphere. Against these harmful effects must be balanced the contribution Wairakei has made to the development of the region, the creation of job opportunities in an expanding population, the value to the economy of New Zealand of cheap, reliable, indigenous energy, and the enhancement of national prestige.

The need to control pollution to the fullest extent possible makes it necessary to examine alternative methods of disposal of the waste heat and waste water for future geothermal developments, of which there are a number possible in New Zealand. A technique for overcoming the main problem, waste water disposal, is being investigated both in New Zealand and overseas. This method, involving reinjecting the waste water into the field, looks promising. Likewise, techniques are available for extraction of the waste gases, although their effect on the environment is so small as to make it unlikely that this will be necessary. Adoption of these techniques would mean a small increase in the cost of geothermal energy, but at the same time, production of the energy would be almost completely pollution free. In fact, the pollution from geothermal power projects could be reduced to the discharge of the waste heat – inevitable in any known thermal cycle – into the atmosphere in the form of warm air and water vapor from cooling towers.

Refinement of Policies and Planning

While the investigation and development of the Wairakei project involved a substantially new technology, it was not necessary to build a new organization or derive new organizational techniques to cope with it. The existing organizations of the Ministry of Works and Development and the Department of Scientific and Industrial Research had developed over the years in such a way as to make them basically discipline oriented rather than activity oriented. This enabled them (and still does) to undertake any new activity with a minimum of organizational problems. Organizationally, the main difficulties arose from the continually changing nature of the work, from scientific investigation to engineering development. These were amplified at times by political requirements for haste.

The formation of the GDL, a limited-liability, private company, is a normal organizational procedure used when the government wishes to join with outside interests in activities of mutual benefit.

The administration of future geothermal projects in New Zealand will differ from that used for Wairakei only to the extent that changes are imposed from outside, quite unconnected with the requirements of geothermal work. These include changes in government's overall energy policy, changes in energy usage patterns, and changes introduced by new and amended legislation covering the protection of the environment, and the rights of people affected by the development of national resources.

The decision to go ahead with Wairakei was made on the basis of the best information available. It was innovative, however, and more than a little faith was required in making that decision. While decisions regarding future development of geothermal energy may still require some faith, today's administrators, with Wairakei's example of cheap, reliable power before them, can make their decisions with more confidence. Wairakei's example has in fact been used in many decisions to develop geothermal energy in other countries.

As the first project of its type, Wairakei had to cope with many new technical problems. Techniques and equipment had to be developed to handle very large quantities of steam and hot water safely. Techniques for geophysical exploration, for material selection, and for geothermal work had to be developed.

Indeed, the need for an understanding of the chemistry of geothermal fields has led to an intensive investigation into the high-temperature chemistry of a number of chemicals, including silica and carbon dioxide. Thus, technically, Wairakei has had, and continues to have, a strong influence on the development of geothermal energy in general, and of hot water fields in particular, not only in New Zealand, but around the world as well.

This influence has been strengthened by the adoption of a policy of complete freedom of access to the information compiled over the years from Wairakei. This policy takes many forms, including making data available to researchers, providing direct assistance through bilateral aid agreements, multilateral agencies, through direct government-to-government approaches, and providing on-the-job training for overseas people not only at the technical level, but in all professional disciplines.

NOTES

(1) "State Hydro-Electric Department. Statement by the Hon. R. Semple, Minister in Charge of the State Hydro-Electric Department," New Zealand Parliamentary Paper D-4, 1947, p. 2.

(2) "State Hydro-Electric Department. Statement by the Hon. R. Semple, Minister in Charge of the State Hydro-Electric Department," New Zealand Parliamentary Paper D-4, 1948, pp. 1-4.

(3) Ibid.

(4) Quoted in memorandum for the Deputy Secretary Department of Scientific and Industrial Research, from the Geological Survey Office, 2nd August 1950.

(5) Ibid.

(6) Rt. Hon. Sidney Holland, Budget speech to the New Zealand House of Representatives, July 22, 1954 (Wellington: Government Printer).

(7) Editorial, Daily Telegraph, Rotorua, New Zealand, July 28, 1954.

3 Hawaii Geothermal Project: United States

Louis J. Goodman
Tetsuo Miyabara
Barbara Yount

PROJECT BACKGROUND

Hawaii, one of the 50 states of the United States, is an island chain located in the central Pacific Ocean, approximately 2,500 miles (4,000 kilometers) west of the continental United States. It is composed of numerous islands extending across 2,000 miles (3,218 kilometers) of the Pacific and is dominated by five major islands (see fig. 3.1). At the northern end of the chain lies the capital and population center of the state, the island of Oahu. Oahu has a population of 718,400, while the state of Hawaii as a whole has a total population of 886,400. Approximately 200 miles (320 kilometers) south of Oahu is the island of Hawaii, which is the youngest and the only volcanically active of the major islands. It has a population of 76,400.

The state as a whole is blessed with magnificent mountains, beautiful beaches, cool valleys, lush vegetation, fertile plains, abundant sunshine, and plentiful rainfall, all of which have made it possible to develop both a thriving tourist industry and a profitable agro-industry in sugar and pineapple. Ironically, the same geography and geologic characteristics that have helped these industries to thrive have also deprived Hawaii of the conventional sources of fuel needed to power them. Because of the islands' recent volcanic origin, no indigenous fossil fuel reserves exist. In addition, geography has isolated Hawaii from potential energy sources. Unlike the rest of the United States, no coal comes into the state by rail, no natural gas is received by pipeline, and no regional grid of electricity serves Hawaii. The only presently feasible source of fuel is oil, shipped in by large tankers. In short, Hawaii is almost totally dependent upon outside shipments of oil, and is thus vulnerable both to disruptions in delivery and to fluctuations in the global market. The overall vulnerability was succinctly summarized by the opening statement from the Mayor's Energy Resources Committee: "In Hawaii, 'Energy is Oil.' We are totally dependent on oil and gas for our energy consumption needs."(1)

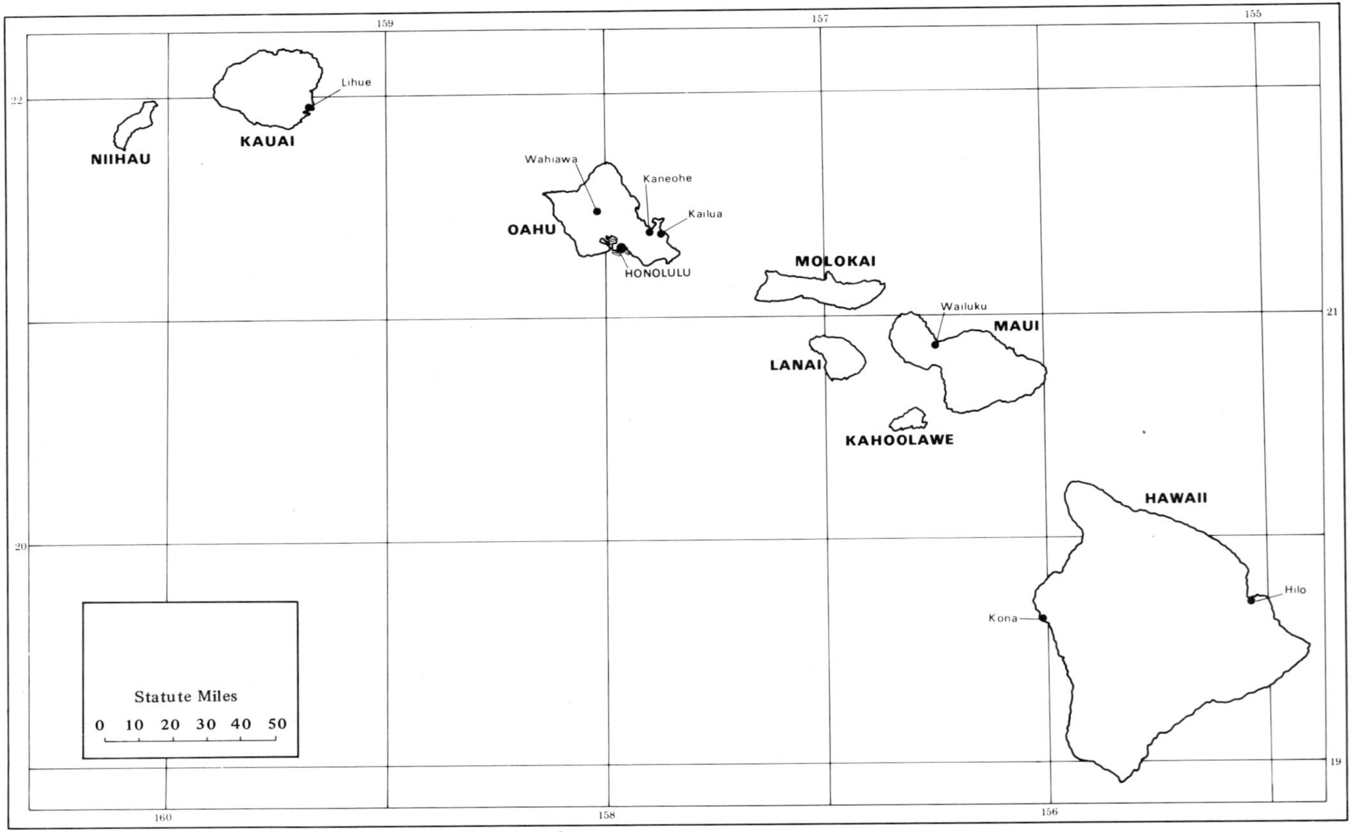

Fig. 3.1. Map of Hawaii.

This vulnerability is underscored by several statistics. From 1960 to 1974, over 99 percent of Hawaii's power was generated from crude oil which was shipped into the state.(2) Approximately 25 percent of all petroleum products was used to generate electricity, 25 percent was used for air transportation, 28 percent was used by gasoline service stations, 14 percent was used for industrial and commercial purposes, and 8 percent was used for a variety of other purposes.(3) Several trends have also increased Hawaii's vulnerability. The cost of oil has risen dramatically; in 1970 it was US$2.50 a barrel and by 1974 it was US$10 a barrel. Between 1958 and 1975 consumption of petroleum rose 300 percent, from 12 million to about 38 million barrels per year.(4)

Yet this dependence and vulnerability upon imported oil is a paradox because Hawaii possesses abundant alternative energy sources – solar, wind, wave, biomass, ocean thermal, and geothermal. Unfortunately, these sources, until recently, have not been developed. A significant factor in this lack of development was the lack of an overall United States energy policy. In its absence, the responsibility for developing alternative energy sources fell both to the individual private utility companies, which generate most of the power in the United States, and to the individual state and county governments. For the utility companies, however, there was little incentive to risk their capital in research and development. Fossil fuel sources, particularly oil, were abundantly cheap, and therefore cost-effective; nonconventional power sources could not compete with oil. Thus, the initiative for developing energy alternatives was left primarily to state and local organizations, such as universities and planning boards. For the most part, these organizations did not have the resources to conduct the research and development necessary to make feasible the various energy alternatives. Nor did they receive necessary support from the state governments, since energy development was not seen as a high priority.

In contrast, the government of the state of Hawaii and the county governments of each island placed a high priority on the development of energy alternatives. Hawaii's vulnerability to oil shortages and dependence on oil had made state and county officials keenly aware of the urgency to develop Hawaii's indigenous sources of energy. As early as 1970, three years prior to the Arab oil embargo, the Hawaii State Legislature passed a resolution requesting the University of Hawaii's Center for Engineering Research to submit a study on the potential of new energy sources for Hawaii. The report was completed in 1971 and it listed a number of alternatives.(5)

Geothermal energy was considered one of the more exciting new possibilities. The concept itself – that of using the heat of the earth to generate electricity – could be applied in Hawaii by harnessing and controlling the sometimes destructive heat of the volcanoes. Since the island of Hawaii was formed by the largest volcanic mass in the world (see fig. 3.2), geothermal energy had great potential. And this potential captured the imagination of many of the state legislators and helped gain local momentum for geothermal research.

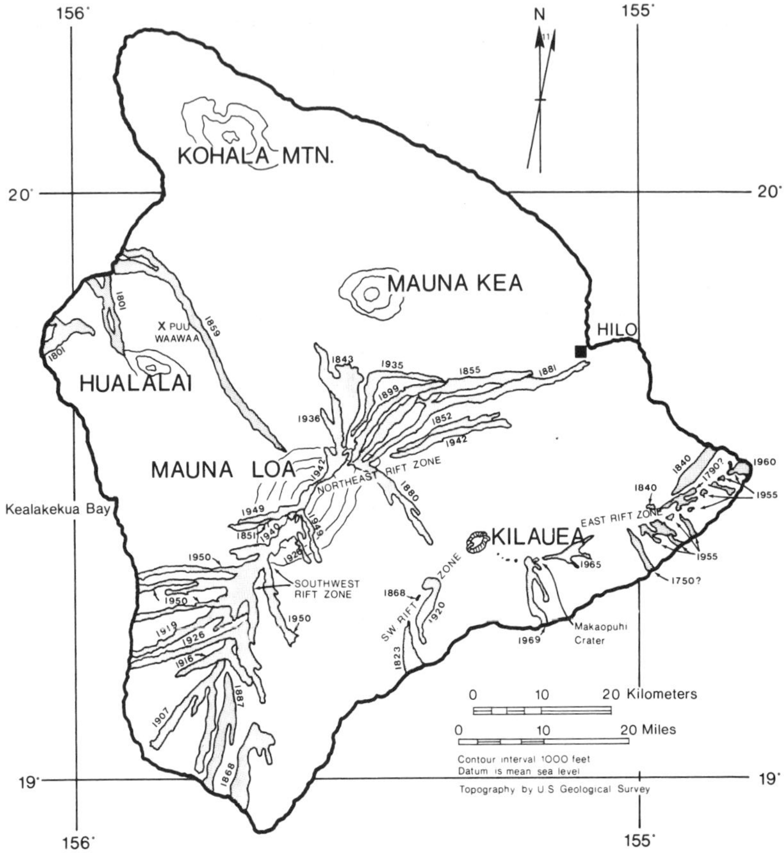

Fig. 3.2. The island of Hawaii with volcanoes.

Additionally, the county government of the island of Hawaii was enthusiastic to make use of its untapped resource. In this regard, several early experiments had been conducted on the island. In the 1920s someone had attempted to use the volcanic heat directly to generate electricity. More recently, in the 1960s explorations for geothermal reservoirs had been conducted. These explorations resulted in the drilling of four wells in the Puna region, the deepest of which was about 700 feet (213 meters). However, no reservoirs were found and the drilling projects were abandoned as economically unfeasible. If any reservoirs existed, they existed at considerably greater depths and could be exploited only at great cost.

PHASE I: PLANNING, APPRAISAL, AND DESIGN

Identification

In 1972, with the encouragement of state and county governments, Howard Harrenstein, the director of the University of Hawaii's Center for Engineering Research and a long-time believer in Hawaii's geothermal potential, submitted a US$2.7 million research proposal to the National Science Foundation (NSF). The proposal outlined a two-year project, called "Project Pele." It was essentially a multidisciplinary effort that would include:

1. Geophysical program: geophysical explorations on the island of Hawaii; these would include surface studies, as well as the test drilling of a series of shallow wells and a deep well, possibly to depths of 10,000 feet (3,048 meters);
2. Engineering program: engineering research into problems associated with geothermal generation of electricity; and
3. Environmental/socioeconomic program: investigations on geothermal energy's socioeconomic and environmental implications.

The project appeared to have one major drawback: although it outlined a very ambitious plan that would result in potentially useful scientific information, it was a pure research project. No plans were made to actually convert geothermal energy into electricity. Nonetheless, the Hawaii State Legislature and the county government of the island of Hawaii strongly supported the proposal, each granting the project $100,000 contingent upon its receiving NSF matching funds. However, the project was not immediately funded.

Instead, in June 1972, NSF awarded a smaller geothermal research grant of about $400,000 to George Keller, a professor of geophysics at the Colorado School of Mines. Keller intended to drill a well about 3,500 feet (1,067 meters) deep in the Hawaii Volcanoes National Park, at a site where the United States Geological Survey indicated there might be underground steam. Even if Keller discovered no reservoir of steam, he would take core samples and conduct geophysical tests. Since these tests would indicate the potential of usable geothermal energy in Hawaii, they would have a strong bearing on the more elaborate Project Pele proposed by Harrenstein. Positive findings would add justification for funding the multimillion dollar research project.

Keller's project was eventually successful in obtaining research data and in drilling to its target depth of 3,500 feet (1,067 meters). The project also encountered certain problems.

One significant problem arose just prior to the drilling. The Congress of Hawaiian People, a group representing the interests of native Hawaiians, asked Hawaii Volcanoes National Park officials to delay the project. A co-chairman of the congress said his group had two objections. First, the drilling might violate religious and spiritual beliefs of the Hawaiian people. He stated that "Hawaiians should have been consulted before the drilling was approved because it will take place on the sacred religious grounds of our ancestors."(6) He argued that Keller should have prepared an environmental impact statement that detailed the site's religious and historical value. Second, he claimed that if the drilling found any commercially usable steam, Hawaiians should profit from it because the state constitution declared that indigenous natural resources should be used for the betterment of the native Hawaiian peoples.

Both objections were eventually resolved. Keller met with the congress and ensured them that the specific project site was not in an area of historical or religious significance. He also obtained a ruling from the Hawaii Volcanoes National Park superintendent stating that any geothermal steam from the site would never be exploited commercially. The park was a public reservation and therefore none of its resources could be bought or sold.

Formulation: The Initial Proposal

While Keller's project was taking place, the proposed Project Pele suffered a setback when Harrenstein resigned his University of Hawaii post to accept another position at the University of Miami. Since Harrenstein was listed as the project's principal investigator, the proposal had to be revised and then resubmitted to NSF. Initial planning meetings were held and the first decision made was to name a management team to replace Harrenstein. The management team was to consist of John Shupe, dean of the University of Hawaii's College of Engineering, George Woollard, director of the University of Hawaii's Institute of Geophysics, and John Craven, the state of Hawaii marine affairs coordinator and the dean of the University's marine program. The project was also renamed the Hawaii Geothermal Project (HGP).

In August 1972, an NSF official advised the HGP management team that a single person should be assigned to manage the project. The failure to name one well-qualified person to assume overall leadership would endanger potential funding. It was then decided to appoint John Shupe as the principal investigator. He would be the project director, responsible for the project's overall administration and management and, in essence, responsible for the degree of project success or failure. Sharing responsibility with him were to be three co-principal investigators: Augustine Furumoto, a professor of geophysics at the University of Hawaii; Paul Yuen, associate dean and professor of the University of Hawaii College of Engineering; and Robert Kamins, a professor of economics at the University. They were to be the coordinators of the geophysics program, the engineering program, and the environmental/socioeconomic program, respectively. Each would devote half his working time to the project and the rest to his normal university duties.

An HGP executive committee was also established. The executive committee was composed of Shupe, the coordinators of each research program, George Woollard, and John Craven. Although this committee would make no direct decisions, it would play a policy making and planning role. Figure 3.3 illustrates the 1972 organizational chart of HGP.

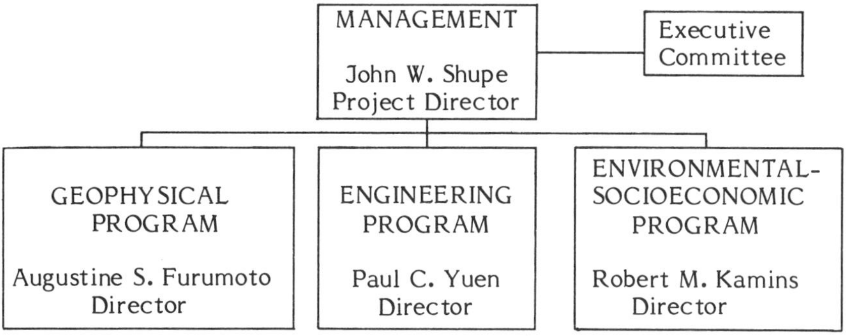

Fig. 3.3. Hawaii Geothermal Project – organizational chart, 1972.

By late 1972, the executive committee realized that the proposed project would have a better chance of being funded if it included research and development that led directly to the conversion of geothermal energy into electricity. This would demonstrate the practical value of the research, as well as provide some tangible results for such costly research.

Consequently, the HGP executive committee expanded the scope of the project to include the planning and construction of a 10-MW prototype geothermal power plant. Like the originally proposed Project Pele, research under GHP was divided into three areas: geophysics, engineering, and environmental/socioeconomic. A total of 37 separate tasks were identified in these three areas (see fig. 3.4) and a team of 54 researchers was named to conduct the investigations.

Requiring US$5 million over a two-year period, the HGP proposal was conceived of in the following three stages:

Stage I was to be the initial, short-range exploratory and applied technology research that would assist in the early development of geothermal power. This stage was intended to acquire scientific information and to help locate possible drill sites for geothermal steam.

During Stage II, the geophysical data collected in the first stage would be used to establish an exploratory drilling program. This stage would culminate in drilling one deep hole that would hopefully tap a steam or hot water reservoir.

Stage III was envisioned as a planning stage for the actual construction of a geothermal power plant. The well would be analyzed and the tests necessary to design the power plant would be conducted. The

Geophysical Surveys	Engineering Studies	Environmental Socioeconomic Studies
Photogeologic survey	Well test analyses	Environmental impact
Aeromagnetic survey	Ghyben-Herzberg lens dynamics	Geotoxicology of thermal areas
Electrical resistivity survey	Geothermal plant optimum design	Land use and land law
Electromagnetic induction survey	Corrosion and wear reduction	Legislation and regulation
Microseismic survey	Electrical energy transmission	Planning
Offshore seismology	Energy extraction from high-temperature brine	Economic analysis
Thermal (300-feet) survey	Materials for use with magma and hot rock	Phytoplankton by-product research
Shallow seismic survey	Direct extraction of magma energy	Agriculture research
Petrology, structural, geology and geochemistry	Direct energy conversion	Trace-metal recovery
Geochemistry of fluids	Alternate modes of energy transmission and storage	
Physical properties of rock	Pilot plant steam production	
Ground water	Mechanical design and layout for pilot plant	
Deep drill	Pilot plant electrical generation and transmission	
Model study		
Evaluation		

Fig. 3.4. Hawaii Geothermal Project – proposed research.

Source: Hawaii Geothermal Project Quarterly Progress Report No. 1, June 1, 1973 through August 31, 1973, pp. 1-3.

actual generator design would then be drawn up. In May 1973, approximately a year after the submission of the project proposal, NSF announced that it was awarding the Hawaii Geothermal Project (HGP) a one-year budget of $252,000. Although other geothermal projects from around the United States had requested funding, it had been decided that a high national priority was to obtain more information on geothermal potential in island volcanic regions. This amount, added to the initial grant of $200,000 which had been given to the project in 1972 by the state and county governments, provided HGP with a first-year budget of $452,000. Although the budget was far less than the originally requested five million dollars, it was viewed as the beginning of a long-term commitment to develop geothermal power in Hawaii. The budget was also sufficient to enable Stage I, the early exploratory research, to get underway.

Feasibility and Appraisal: 1973-1975

The initial research activities of HGP were, in a sense, the formal feasibility studies, determining the chances of the project to eventually generate electricity from Hawaiian geothermal sources. The exploratory geophysical surveys would help determine if and where reservoirs of steam or hot water existed and thus allow NSF to assess HGP's scientific feasibility. The engineering and environmental/socioeconomic studies would identify and clarify the technological, environmental, legal, regulatory, and economic problems that could hinder the eventual development of geothermal power in Hawaii – provided, of course, that a usable source of geothermal energy existed.

Overview of initial project activities

Upon word of NSF's allocation to HGP, the state government and the county government of Hawaii Island, which both strongly supported the development of alternative energy sources, released funds to HGP to begin planning activities. Shupe then convened a meeting of the executive committee and each of the program coordinators was given a separate budget. This would provide each coordinator with the independence and flexibility needed to administer the research in his area and, at the same time, make each accountable for his program. Shupe would be responsible for the overall management and coordination of the project. He would also be responsible for maintaining project cohesiveness and for ensuring that each research program developed consistently with overall project goals.

The executive committee also had to make important budgetary decisions. The separate budgets initially proposed by each program coordinator had to be reduced because the initial level of funding was considerably less than anticipated. Thus, it was decided that each program would receive only enough funds to initiate crucial tasks. The bulk of the money, however, would be allocated to the geophysics program because it was conducting the exploratory surveys crucial to the project's continued progress. Despite the reduced funding, it was anticipated that NSF or some other government agency would provide HGP with additional funds if the geophysical surveys indicated the existence of a reservoir of steam.

With these decisions made, the directors of the geophysics, engineering, and environmental/socioeconomic programs began organizing their respective activities.

Geophysics program

The geophysics program director, Augustine Furumoto, had requested about $800,000 for 1973-1974 activities, but received only about $250,000. He decided to limit the geophysical surveys to those which could begin immediately; the remaining surveys depended upon the data from the initial surveys and would be undertaken if additional funds were received. The chosen surveys were those crucial in identifying a potential reservoir of steam or hot water. Providing clues about the subsurface conditions, they would be like pieces of a large jigsaw puzzle, which, when put together properly, would serve as a geophysical model of the volcanic area. These surveys included: 1) photogeologic, 2) geoelectrical, and 3) microseismic, and geochemical. Other surveys, such as the gravity and magnetic, were planned for 1974, if funds were available. The photogeological survey, contracted out to a commercial firm with both the necessary equipment and experience, involved flying over the volcanoes at approximately 2,100 feet (640 meters) and taking infrared photographs of the rift zones – zones of innumerable fissures that served as underground pathways for the rise of magma. When developed, the photographs would expose gradations of surface heat and would locate volcanic vents and other "hot spots" along the rifts. Any surface temperatures that exceeded the highest range of the film would be exposed as spots of white. Flights were conducted during August 1973 and the photos were developed soon afterwards. The photographs revealed a concentration of white dots along the east rift of Kilauea volcano, in an area named Puna. The temperature range indicated by the film was 61°F (16°C) to 77°F (25°C).

The initial electrical resistivity survey was subcontracted to George Keller, who already had equipment in the field from his earlier project. These surveys, called dipole-bipole mapping, checked the earth's electrical resistance by passing an electrical current between two poles set in the ground; low resistivities readings indicated conductive foundations, such as hot saline water or highly conductive soils. Since low resistivity readings could also indicate pipes or electrical wires, several surveys had to be conducted to help determine the true sources of the readings. The survey indicated two areas of low resistivity that could be attributed to thermal sources – the Opihikao anomaly and the Pahoa anomaly, both located in the area of Puna. The Opihikao anomaly had resistivities of about 5 ohms per meter (ohm/m) from 1,969 to 6,890 feet (600 to 2,100 meters), while, between the same depths, Pahao had resistivities of about 8 ohm/m.

The microseismic surveys were to have measured the velocity at which sound passed through the ground, and thus provide indications about the area's subsurface structure. These surveys were postponed because of delays in receiving the necessary equipment. The geophysics coordinator agreed, however, to begin a ground noise survey, which would indicate the variance in subsurface noise. Since in Hawaii, sites of volcanic activity produce intense sound, these surveys would help locate sources of geothermal heat. Like the initial surveys, the ground noise survey discovered intense sound.

The geophysical surveys progressed steadily through early 1974, with the geophysics coordinator assigning tasks to appropriate geophysicists on the project team. Often, the assigned individual would subcontract a specially equipped commercial firm to conduct the survey, and then analyze the data himself. This arrangement proved satisfactory for many of the tasks. However, when no commercial firm could undertake the surveys, the geophysicist would have to order special equipment or redesign existing equipment, and then conduct the survey. This led to some delays and the surveys fell behind schedule.

Engineering program

The engineering program, like the other two research programs, had to reduce the number and scope of its initially proposed research tasks. The engineering director, Paul Yuen, thus decided to concentrate on: 1) geothermal reservoir engineering, and 2) optimal geothermal plant design. These two tasks dealt directly with applied research crucial to the production of geothermal energy in Hawaii.

Geothermal reservoir engineering was initially two separate tasks, but because of their close linkage, these were later collapsed into one task with two related components. These components included: 1) numerical modeling, and 2) well testing and analysis. The engineers working on numerical modeling attempted to computer-simulate the operational dynamics of a geothermal system under different conditions. To derive the mathematical relationships, they first had to investigate several issues. How, for example, would pumping, reinjecting, and recharging the geothermal well affect the Ghyben-Herzberg lens? The Ghyben-Herzberg lens is a pool of fresh water trapped in porous rocks beneath the island's surface. Sea water also permeated the island's subsurface, but the fresh water was lighter and thus floated on the sea water, forming a lens which supplied much of the island's water needs. When the engineers completed these investigations, they would construct a model that would generate computer answers to questions such as: How deep must the well be drilled to avoid destruction of the Ghyben-Herzberg lens? What is the life span of the well? What is the capacity of the geothermal reservoir?

Well testing and analysis would culminate in engineers and geologists going into the field to test measure the geothermal well – assuming, of course, that a successful well were drilled. The task would proceed in several stages. Initially, the engineering team would evaluate the existing equipment and the methods used by the geothermal engineers in the rest of the world; this involved a literature search. Then they would examine the techniques used by petroleum reservoir engineers to measure oil wells. However, since the volume and capacity of a geothermal reservoir depended on temperature, but a petroleum reservoir did not, many analytic techniques of petroleum reservoir engineering were inadequate. Thus, during the next stage of research, engineers would modify and adapt both the geothermal and the petroleum methods to develop a comprehensive geothermal testing program. The completed program would be appropriate for a geothermal well and would include a complete array of geological and reservoir engineering tests, as well as recommendations for the purchase of equipment. After the researchers developed the well testing program, they would conduct the tests on the well itself. Ultimately, the data collected would help predict the life span and capacity of the geothermal well.

The engineering program's second research task was to study power plant designs that could be used if a geothermal well were discovered. Since the optimal power plant design depended upon the form of energy produced by the well (it might be in the form of dry steam, wet steam, hot water, gases, dissolved solids, or vapor) the engineering team might have to study many options. The team decided, however, to limit their investigations to two basic types of geothermal power plants: the vapor flashing plant and the binary fluid plant.

The vapor flashing plant would be practical if the well produced geothermal steam. In this system, the geothermal well would contain hot water under intense pressure. As the water would rise from the bottom of the well, the pressure on it would decrease and some of it would flash to steam. The well would thus emit a mixture of hot water and steam. The steam would be separated from the hot water by a separator and piped directly to a turbine generator. The hot water could be discarded or could be piped to another separator, which would further reduce the atmospheric pressure on the water, thus causing it to flash to steam. This steam would then be piped to the generator (see fig. 3.5).

The binary fluid plant would be efficient if the well produced hot water. In this system, the hot water would be used to heat a secondary liquid, such as isobutane. When the isobutane became vaporized, it would power a turbine that would then produce electricity (see fig. 3.6).

To design the optimal plant, the engineers would have to answer questions such as: What would be the most efficient steam pressure to power different size plants? What plant configuration would be most feasible given different well conditions? What kind of turbine generator should be used if the well produced wet steam, dry steam? How should the plant's discharge system be designed to make it environmentally sound?

The engineering team began work on each of these tasks in late 1973, and the research proceeded smoothly through 1974. However, much of the research was intended to be applied to the actual

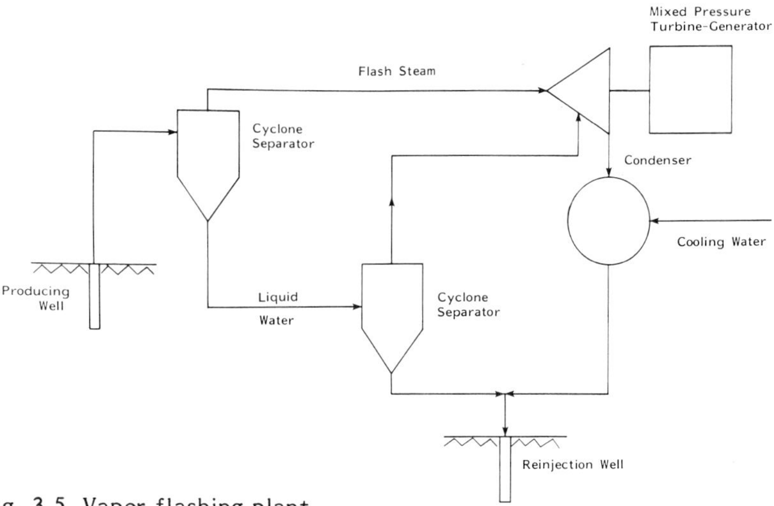

Fig. 3.5. Vapor flashing plant.

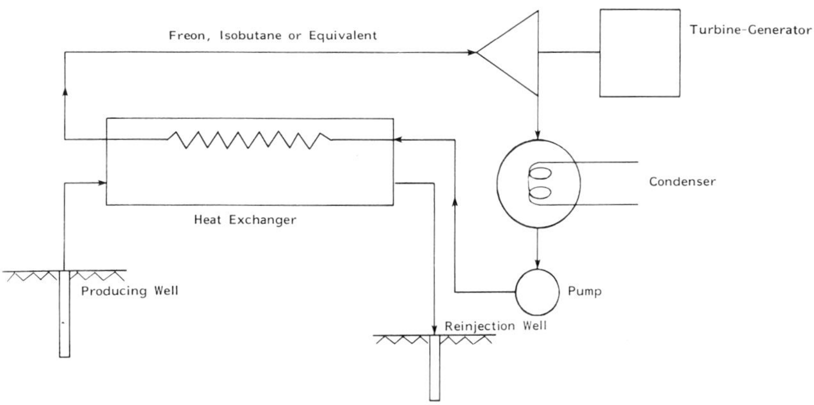

Fig. 3.6. Binary flashing plant.

production of geothermal energy in Hawaii. Thus the engineering program could be successful only if money were obtained to continue the overall project and only if a successful geothermal well were drilled.

Environmental/socioeconomic program

With only limited funding available for environmental and socio-economic studies, the coordinator of the program, Robert Kamins, decided to focus on the following three aspects of geothermal development: 1) legal and regulatory aspects, 2) economic implications, and 3) environmental impacts. These topics had a direct bearing on the social, economic, and political factors that would help or hinder the implementation of geothermal energy in Hawaii.

The first aspect, the legal-regulatory research, involved the complex questions of ownership and government regulation. Because Hawaii public law did not cover geothermal resources, it was uncertain whether geothermal steam was publicly or privately owned. In order to clarify this issue, the research team would first have to examine how other areas in the United States settled the ownership question. These approaches would then have to be compared to the relevant statutes in Hawaii, and alternative solutions proposed that were consistent with Hawaii's statutes. Of particular importance in this regard was whether geothermal resources could be classified as mineral, water, or a substance unique in nature. If geothermal resources were classified as mineral, some would be owned by the state under mineral rights clauses; if they were classified as water, they would be owned by private landowners through legal precedence; and if they were classified as unique in nature, their ownership would be uncertain. This situation was further complicated by two issues. First, some of the land deeds issued during the early 1900s did not reserve to the state of Hawaii the exclusive ownership of any subsurface minerals. Since it was not known how many of these deeds existed or where the land pertaining to them was located, ownership could be determined only by reviewing individual land deeds. Second, several groups representing native Hawaiian rights claimed that the geothermal resources belonged to native Hawaiians and that their ownership was upheld by the state of Hawaii constitution.

In examining the regulatory issues, the environmental/socio-economic team would have to address questions such as: Which government agencies, if any, should possess authority over the drilling, land use, and development of geothermal energy? What safety requirements should be adopted for drilling and for geothermal power plants? What environmental safeguards should be imposed upon geothermal development? How should the public interest be protected?

A fundamental issue regarding these questions was the multiplicity of government agencies potentially involved in regulation. On the federal level these agencies included the Environmental Protection Agency (EPA) and the Energy Research and Development Administration (ERDA). On the state level the agencies were the Department of Health, the Department of Transportation, the Department of Planning and Economic Development, the Department of Land and Natural Resources, the State Energy Resources Coordinator, the Department of the Attorney General, the Office of Environmental Quality Control, the Hawaii State Legislature, the Public Utilities Commission, and the Department of Regulatory Agencies. On the county level the agencies were: the Department of Public Works, the Department of Research and Development, the Department of Water Supply, and the Planning Department.

The second research topic – the economic impact of geothermal development – involved the research team building an econometric model that would provide projections to the year 1980. To begin this task, the researchers would collect data on the cost, source, amount, and distribution of Hawaii's present energy use. Data would also be collected on the cost and production of geothermal energy around the world. From this data, the researchers would make certain assumptions about energy prices. Then, after building a dynamic model, estimate the demand for geothermal energy under varying supplies. Projections could then be made of the resulting impact upon employment, population dispersion, industrial growth, public revenue, and economic growth.

The third research task – environmental analysis – would monitor the ecologic impact of any geothermal well or power plant developed by HGP. Initial studies would involve researchers collecting baseline data of the vegetation and wildlife in the drilling area. Of particular concern, however, was the Ghyben-Herzberg lens, which supplied much of the island of Hawaii's fresh water. A medium or deep geothermal well might penetrate the lens, thus endangering the water supply. To provide information about the impact of drilling, researchers would initially measure nearby springs for salinity, temperature, and chemical characteristics. Then they would establish a program to monitor the springs for any deviations from the baseline measures. Eventually, the environmental program would complete a comprehensive environmental impact statement.

In late 1973, the program team began working on the legal/regulatory aspects and the economic implications. The environmental analysis, however, could not be initiated until a potential drilling area was designated, and in late 1973, it was uncertain if HGP would receive drilling funds. Nonetheless, there was an urgency for the study on legal implications because the Hawaii State Legislature was considering legislation to clarify the ownership of geothermal resources.

To assist the legislators, the research team completed, in February 1974, a preliminary analysis of all geothermal ownership options and their consequences. Aided by this study, the legislature passed the state's first geothermal law. It classified geothermal resources as mineral, thus reserving them to the state under mineral rights provisions. Work on the regulatory aspects and the economic impact continued through 1974.

Overall Coordination and Management of HGP 1973-1974

While each program's research was being conducted, John Shupe attempted to ensure continued support for HGP and to maintain the project's overall cohesion. He conferred regularly with federal, state, and country agencies, made presentations of the envisioned HGP program at public symposiums and at international conferences, and formed contacts with a widespread network of geothermal experts, who would provide HGP with advice, information, and assistance. In August 1973, Shupe also formed the Hawaii Advisory Committee (HAC) and the National Liaison Board (NLB).

The HAC was composed of Hawaii's business, political, and community leaders such as the president of Hawaii's major electric company, the director of the state Office of Environmental Quality, the director of the state Department of Planning and Economic Development, the director of a leading environmental group, the president of the Congress of Hawaiian People, and officials from the County of Hawaii. Since these individuals represented the groups that formulated Hawaii's energy policy, their support was critical for the successful development of geothermal energy. Moreover, many of the groups, particularly the state and county officials, strongly supported the development of indigenous alternative energy sources. Therefore, it was natural that the project include them in an overall cooperative effort. The first HAC meeting was held in October 1973, and the group decided to meet semiannually.

The NLB was composed of geologists, geophysicists, and engineers from the U.S. mainland. Experts on geothermal power development, they would monitor and advise HGP on its progress and direction. Since they also worked for key agencies such as the National Science Foundation (NSF) and the United States Geological Survey, they would be extremely influential in ensuring continued federal funding for the project. The first NLB meeting was scheduled for early 1974.

Maintaining the overall cohesion of HGP was the project director's most difficult but most important task. He realized that the project was, to a certain extent, naturally segmented because each program conducted very different kinds of research. Moreover, some separation was necessary to give each program the flexibility and independence necessary to accomplish its individual goals. At the same time, however, he also realized that the ultimate goal of each research program was to support the generation of geothermal power in Hawaii. This goal provided the driving force for HGP and integrated the research programs with one another. Thus, the project director had to encourage all program coordinators to keep the ultimate goal in mind and to avoid concentrating on research that was not relevant to the project's overall goals.

Critical Events and Decisions of 1974

At the beginning of 1974, HGP was increasingly embroiled in the major policy question of whether or not to establish and proceed with an experimental drilling program. The project director felt that the project had to make progress by moving in the direction of drilling an experimental well. He thus advocated the establishment of a drilling program. The geophysics program coordinator felt that further research had to be conducted before HGP could even consider a drilling program. However, the engineering and socioeconomic program coordinators supported the establishment of a drilling program. Other HGP team members, particularly Agatin Abbott, a professor and the chairman of the Department of Geology and Geophysics at the University of Hawaii, supported a drilling program because it was the only way to actually determine if a usable geothermal source existed. However, a final decision could not be made immediately.

Against the background of this unresolved policy issue, Shupe convened the first meeting of the National Liaison Board. The meeting was held in February 1974 on the island of Hawaii, where any potential drilling would occur. The meeting was intensive. The director of the NSF's Advanced Energy Research Program outlined the foundation's interest in HGP, emphasized the crucial role of the NLB in evaluating HGP's progress, and pointed out that NSF could not fund commercial exploratory drilling but could fund a research drilling program.

Then, each HGP program coordinator presented a progress report. The engineering coordinator described the reservoir engineering and mathematical work to date; the environmental/socioeconomic coordinator described the Hawaii State Legislature's efforts to establish a legal framework for the ownership of geothermal resources. Most of the meeting was spent, however, on the progress of the geophysical program. Initially, the geophysics coordinator described the infrared air photo survey and the electrical resistivity surveys and also presented data from his deep (4,000 feet or 1,257 meters) drill. A review of surveys conducted prior to the formation of HGP was also presented and a lively discussion ensued. Board members, HGP personnel, and persons in the audience asked probing questions and offered interpretations of the geophysical data.

At the end of the meeting, the NLB could reach no consensus on which sites had greatest geothermal potential. But it did agree on recommended courses of action for HGP. The NLB felt that HGP should move rapidly on establishing a research drilling program. There was no other way to test the theories and interpretations. The NLB further recommended that the coordinator of the drilling program be Agatin Abbott. Abbott was the senior geologist on the research team and he had conducted the aerial infrared surveys. He also was an advocate of an early drilling program and had vigorously supported its establishment. NLB also advised that a site selection committee be formed. This committee should be composed of senior geologists and geophysicists, who would collectively make decisions about all aspects of the drilling program, including the number of research wells and the location of all drill sites. Abbott was recommended to chair the site selection committee. After reaching a consensus on these recommendations, the board members concluded the first NLB meeting.

Shupe informed the Hawaii Advisory Committee about the National Liaison Board's recommendations and the HAC likewise encouraged the organization of a drilling program. The HAC also told Shupe that they would aid HGP in securing state government funds to support the drilling. With the pledge of HAC's support, it was decided to request funds from the Hawaii State Legislature, which was just beginning its 1974 session. In this process, a lawyer first drafted an appropriations bill requesting $500,000 from the state of Hawaii government. The bill was then introduced to the legislature and each program coordinator testified before the legislature's appropriations committees about his program's progress and about its relationship to the development of geothermal energy in Hawaii. The keystone of the testimonies was a discussion of geothermal drilling by Abbott. Since a drilling program had not yet been formulated, he could provide only a general overview. Nonetheless, the overview was sufficiently detailed to capture the legislators' interest. At the conclusion of the formal testimony, Shupe was assured that HGP would receive support. But he would have to wait until the legislature formally approved the funds and until the governor released them.

In late February 1974, the project director determined that it was timely for the HGP to formally establish a drilling program. Thus, the HGP executive committee was convened to discuss the issue and reach a decision. At the meeting, Augustine Furumoto, the geophysics coordinator, stated that establishing a drilling program was premature; HGP funds could be most usefully spent for further geophysical surveys and data interpretation. In response, Shupe argued that both the NLB and the HAC, the two advisory groups, had strongly recommended the formulation of a drilling program. Moreover, it was an appropriate time to establish and proceed with such a program because the project had to progress toward its long range goal of generating geothermal power.

Paul Yuen, coordinator of the engineering program, Robert Kamins, coordinator of the environmental/socioeconomic program, and other members of the executive committee also pointed out that the geophysical surveys had generated a large quantity of data. The data had been interpreted and varying predictions of the subsurface conditions had been made. But the only way to check the interpretations and to verify the accuracy of the predictions was to actually drill. Finally, the drilling proponents argued, the area in which the drilling would occur was one of the most thoroughly studied geological regions in the world. The HGP, the University of Hawaii, the United States Geological Survey, the state and county governments, and other universities had conducted geological and other survey expeditions in the area. Surveys such as geodetic, gravity, deformation, seismic refraction, magnetic, and thermal had already been conducted. An experimental drilling program was thus long overdue.

After discussing the issue for a while longer, the executive committee voted to establish a drilling program. Abbott was appointed the program coordinator, and Shupe provided him with a small budget to initiate planning activities. Abbott's first act as coordinator was to form a site selection committee, which would assist in planning and in making decisions for the program. By March 1974, then, HGP consisted of four research programs with several advisory and policy boards. (HGP's organizational structure as of April 1974 is illustrated in fig. 3.7.)

Following these actions, the project director turned his attention to fiscal management. He realized that HGP would require more time and more money to complete the research programs presently under way. Additionally, a large grant was required to support the research drilling program. It was therefore decided that each program coordinator would assist in the preparation of: 1) an eight-month budget extension to fund their current activities, and 2) a new proposal to fund an experimental drilling program, including related activities.

Design: Completing the Proposals and the Drilling Program

From March through June 1974, each research team continued to make progress on its scheduled research tasks. The engineering program's researchers continued working on the numerical modeling and the well testing program. Additionally, they continued designing the optimal geothermal plant. In the environmental/socioeconomic program, research continued on the legal-regulatory issues and the economic analysis.

Finally, in the pivotal geophysics program, work was being done on a wide variety of exploratory surveys. The geophysics researchers completed building a wire loop magnetic induction system, and implanted 12 pairs of electrodes deep into the earth by air-dropping them in inert missiles. This system was designed to follow up on the initial electrical surveys and would help to determine whether or not geothermal resources existed at depths of 1.24 miles (2 kilometers). Other surveys that continued under the auspices of the geophysics program included a microseismic survey, a magnetic survey, and geochemical surveys of ground water. Also, during this time period Donald Zablocki of the United States Geological Survey, with the assistance of HGP team members, began an electrical self-potential survey of the Puna area. Self-potential surveys had, in the past, proved extremely useful in helping locate potential geothermal sites.

Important as the geophysical research was, however, the top priority of HGP increasingly shifted to completing the budget extension and the experimental drilling proposal. Preparing the budget extension was relatively simple but time consuming. Each program coordinator wrote a progress report elaborating on his program's research results, on the problems encountered, and on the funding necessary to complete the research. The project director reported on the overall status of the

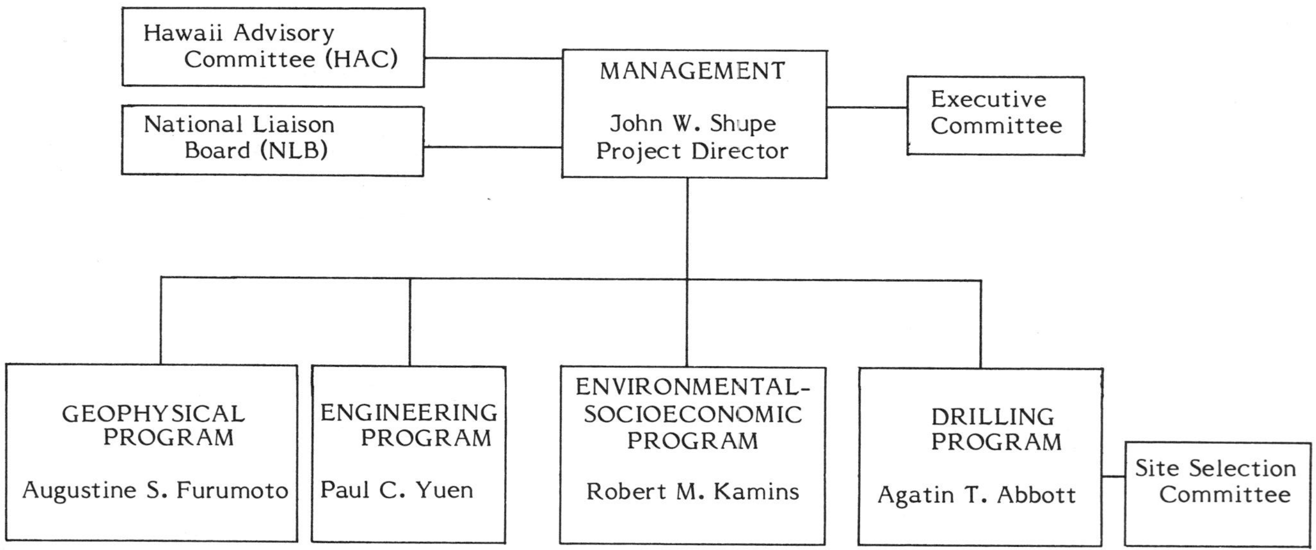

Fig. 3.7. Hawaii Geothermal Project – organizational chart, 1974.

project and justified the requests for a time extension and for additional funds. When completed, the 108-page budget extension requested an additional $340,000 extended over an eight-month period ending in December 1974. It was submitted to NSF in April 1974.

The executive committee then met to discuss the preparation of the experimental drilling proposal. The geophysics coordinator emphasized that the tone of the proposal should stress the project's continuing experimental focus. The intent of the drilling program would thus be to check the geophysical predictions and interpretations – not to discover a usable geothermal resource. In response, the project director and the other program coordinators acknowledged the experimental nature of the drilling, but added that the ultimate goal of the project was to develop the capability for generating geothermal power. This meant that it was crucial to discover a usable geothermal source. They thus emphasized that the proposal should retain a judicious balance between pure research and the development of practical applied techniques.

After discussing this issue, they decided that a sharp restatement of the overall goals of HGP was necessary to keep the drilling proposal in perspective. These goals included:

1. Improvement of geophysical survey techniques for locating underground heat resources.

2. Identification of potential geothermal resources, initially on the Big Island. . . .

3. Experimentation with deep-drilling techniques for subsurface heat.

4. Development of efficient, environmentally clean systems for conversion of underground heat resources to useful energy.

5. Completion of socio-economic and legal studies for conversion of underground heat resources to useful energy.

6. Establishment of environmental baselines with which to monitor subsequent geothermal development.

7. Development of a geothermal production field and prototype power plant on the Big Island. . . .(7)

With these goals in mind, all agreed that the proposal would request funds for each program to continue its research tasks. However, the priority research tasks would be those that directly supported or related to the drilling program. Finally, they agreed that the proposal would give highest priority to the drilling program itself.

The task of writing up the proposal was divided among the program coordinators; each coordinator would be responsible for the write-up of his program's plans. The project director would coordinate work on the proposal, assemble it, make it cohesive, and compile all of its financial and administrative sections. The program coordinators completed the plans for their programs in May, and, by late June 1974, it was completed. It was divided into four sections; a brief description of each section follows:

Geophysics program. During the time period of the proposal, work would continue on all geophysical surveys that were not yet completed; these would probably be the microseismic studies, the magnetic studies, and the geochemical surveys. Additionally, follow-up studies would begin; these would include a geochemical survey, thermal surveys of well water, and mathematical modeling. Also, in early 1975, the geophysics team would make a preliminary analysis and interpretation of the data to help the drilling program determine the most useful sites for drilling. Finally, two new tasks would be undertaken: 1) a hydrology study of the Puna area, and 2) a study of the physical property of rocks in the same area. The hydrology study would analyze geochemical data to determine the source of geothermal water, the way it circulated beneath the earth, and the process in which it recharged the geothermal reservoir. The physical property of rocks would measure the thermal conductivity of the drilling area.

Engineering program. During the grant period, research would continue on the optimal geothermal plant design and the numerical modeling. However, the priority task would be carrying out the well testing program, which the engineering team had designed. The testing program was composed of three sections: 1) bore hole tests, 2) well completion methods, and 3) well tests.

1. Bore hole tests would be conducted during the drilling. Researchers would continuously monitor the temperature and composition of geothermal fluid; and at regular intervals, as well as periods of sharp temperature increases, they would take formation logs. Formation logs would include information such as temperature, pressure, and composition of the fluid in the drill hole, as well as the type, density, and porosity of the rock surrounding the drill hole. The engineering researchers would obtain this information by drawing core samples of the earth and by lowering a probe into the drill hole. The bore hole tests would not only provide valuable data, but also would help determine how deep to drill and when to stop.
2. Well completion involved deciding how to complete the drill hole and what kind of equipment to use to build the well head. If the well tapped a favorable geothermal reservoir, the drill hole would be prepared for further testing by installing a slotted liner, or a gravel pack. After the hole was completed, the well head would be assembled. The specific equipment to complete the well head would be chosen after further study. But the well head design required: a) a valve assembly to control the flow of steam from the well, b) a silencer to reduce the roar of the steam as it flashed from the well,

and c) a centrifugal cyclone separator to separate the well's steam and hot water. The final task of well completion would be starting the well. If the well did not flow naturally, the engineering team planned to force the geothermal fluid to the surface by injecting compressed air into the hole; thereafter the natural pressure and heat of the geothermal reservoir would force the steam and water up the drill hole to the well head.

3. Well testing would occur in two stages, a downhole fluid measurement stage and a well-flow stage. In the downhole stage, the engineering team would lower measuring probes into the well and record the temperature, pressure, and flow rate of the geothermal fluid. This information would help the engineers to estimate the life span and generating capacity of the geothermal reservoir. After allowing the subsurface conditions to stabilize for at least one month, the engineers would initiate the flow testing. The steam would then be allowed to flow out of the well for extended periods, during which the engineers would measure the pressure, temperature, and mass flow rate. This data would enable them to estimate the well's generating capacity.

Environmental/socioeconomic program. During the grant period, the researchers on this program would complete the legal-regulatory studies and the economic studies. They would also begin two new tasks, environmental monitoring and land use studies. The environmental monitoring would include a baseline data collection of the chemical, biological, and physical characteristics of the area. This baseline would establish the standard against which to measure the impact of drilling. If the investigations indicated that the environment would be adversely affected, drilling plans would be altered. Finally, the fumes produced by the discharge of geothermal steam would contain gases such as ammonia and hydrogen sulfide. To measure the quantities of gas being released into the atmosphere, a special team of scientists would conduct air quality studies. The land use studies would provide two crucial bits of information. First, they would provide information on the zoning codes and the land use laws that might restrict the well location and drilling operation. Second, the studies would identify the owners of potential drill sites. Once this information was compiled the team members would negotiate with the owners for the rights to enter their land and to drill for geothermal resources.

The drilling program. This was the focal point of the proposal. Not only would drilling demonstrate the success of HGP's initial research efforts, but also it would provide dramatic evidence of HGP's progress toward generating geothermal power. Based on the work of each research program, as well as previous geological and ground water surveys, the proposal envisioned drilling experimental holes in three general locations. The most favorable location and the one where drilling would first take place was along the east rift of Kilauea near Puna. In this area, three types of holes would be drilled. These were:

1. Shallow holes (average depth 500 feet or 152 meters) for water samples and temperature measurements;
2. Intermediate-depth holes (2,000 feet or 610 meters) for temperature measurements, rock alternation, water chemistry; and
3. Deep hole (6,000 feet or 1,829 meters or more) to try to reach a potential geothermal source.

To manage all drilling operations, the site selection committee would contract an experienced geothermal engineering firm. This firm would be responsible for overall drilling management, including: drawing up a drilling contract, subcontracting the drilling, managing the drilling operations, drawing up safety regulations, cleaning up the site, controlling the finances, and handling all other operational aspects. The site selection committee would decide upon the location, number, diameter, and depth of all drill holes. Additionally, the committee – in conjunction with the engineering, geophysics, and environmental/socioeconomic programs – would determine the types of scientific measurements, when to take them, and how to assess the results. The site selection committee would also hire the geothermal engineering firm.

In the description of the drilling plans, no specific drill sites were identified. Instead, the plans provided detailed maps of the general areas being considered for drilling and provided the geophysical data which indicated that geothermal resources existed in these areas. The site selection committee would choose specific sites after the results of other geophysical surveys were received from the field and after the socioeconomic program indicated which areas.

The proposal was submitted to the National Science Foundation in July 1974 by the project director, John Shupe. The proposal established a one-year activity period, January-December 1975, and requested $2,000,000, of which $1,200,000 would be allocated to the drilling program.

PHASE 2: SELECTION, APPROVAL, AND ACTIVATION

Selection and Approval

Initial success and some revisions

In May 1974, formal notice was received that the Hawaii State Legislature had approved the Hawaii Geothermal Project's request for drilling funds; HGP would be allocated $500,000, provided that the project also received federal matching funds. Later in the month, the National Science Foundation (NSF) approved the budget extension. NSF would grant the project $217,000 to enable the research teams to continue working on their tasks through December 1974. The NSF program manager further informed HGP that it would receive an additional $118,000 in 1975 to complete the research.

In July 1974, as previously discussed, Shupe submitted the multi-million-dollar experimental drilling proposal to NSF. Since the amount requested was so large, it would take six months to review. While the proposal was being reviewed, each research team continued with its research tasks. Then in September 1974, Shupe conferred with the NSF program managers about the proposal. At this conference they informed him that NSF would not fund such an expensive drilling program, but that HGP could expect to receive approximately $500,000, which was enough to drill a single deep well. Moreover, HGP should specify a drill site.

Shupe informed the HGP executive committee of the funding constraints and then told the program coordinators that they had to revise their programs. The revisions were to include: the geophysical evidence and analysis that had been completed since the submission of the initial proposal, reductions in the level of spending and activity of each program, new drilling plans based upon the new budget constraints, and specific locations of each drill site. Each research team proceeded with the revisions; the bulk of the work, however, was to be done by the geophysics team and the drilling program team in conjunction with the site selection committee.

The site selection committee met in October 1974 to draw up new drilling plans. After discussing all options and reviewing all the available data, they made three decisions. First, they decided that, rather than spend the funds on several shallow holes to gather more information, they would drill <u>one</u> deep hole to possibly tap a geothermal source. Second, because funds were so limited, the committee decided that no engineering firm would be hired to supervise drilling operations. Instead, a drilling consultant would be contracted, and the drilling program team would manage the drilling. Third, the committee decided upon the single drill site.

The drill site decision was critical and there was considerable pressure on the committee. Although the drill was explicitly intended to increase scientific knowledge, it was implicitly intended to tap an as yet hypothetical geothermal source. Thus, since only one hole would be drilled, the committee had to select a site that overlay a geothermal reservoir. Adding to the pressure on the committee was scientific uncertainty. The evidence from the geophysical surveys suggested underground hot water, but was not definitive. Several exploratory holes are normally drilled because, as one expert had commented to the committee, "the odds of finding a usable steam reservoir in drilling are one in five, and geothermal search requires a good deal of luck."(8) Nonetheless, the committee had to select the site.

It considered two general locations, Area A, which was the Pahoa anomaly identified by the electrical resistivity survey and Area B, which was the Opikahi anomaly also identified by that survey. Both areas were situated along the east rift of Kilauea near Puna. Area A was about 1,500 feet (457 meters) north of the Puulena Crater; and Area B was located approximately 3 miles (4.8 kilometers) west (see fig. 3.8). After deliberating, the committee chose Area A and desig-

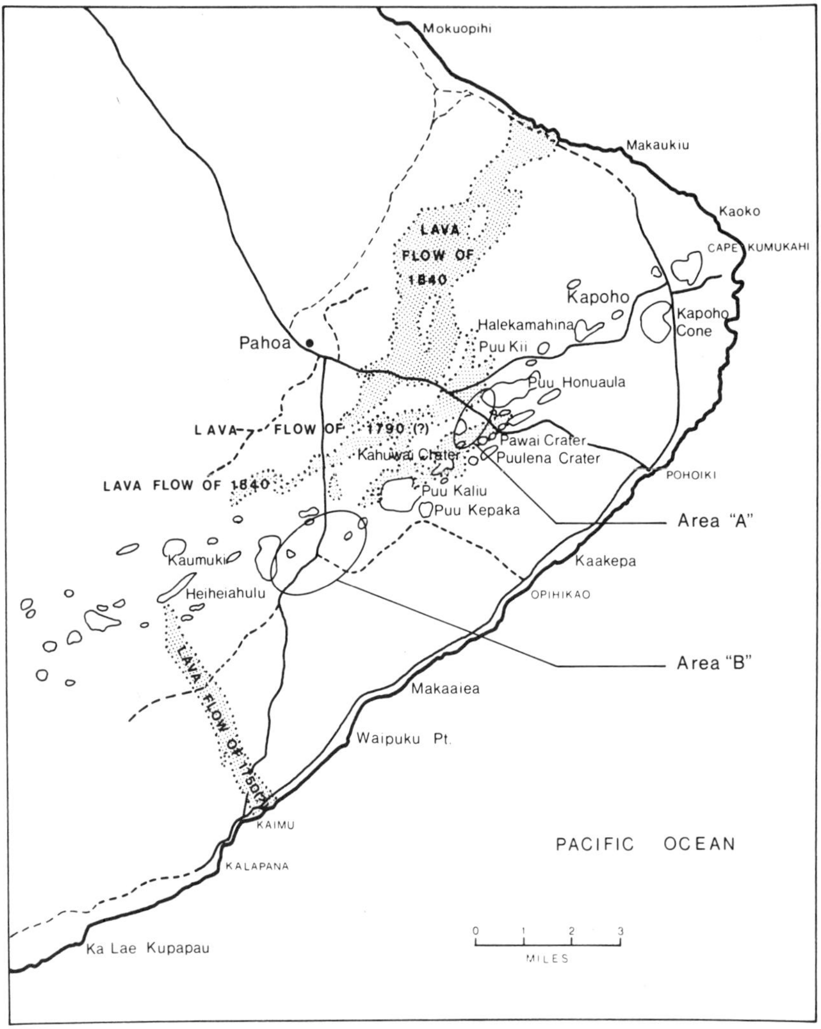

Fig. 3.8. Locations of Area A and Area B.

nated a drill site at the apex of the anomaly. An alternative site was also selected approximately 1,500 feet (457 meters) north of the apex.

In choosing Area A, the committee relied primarily on geologic conditions and the self-potential survey. The geological conditions of the region included a history of volcanic activity, an interesting offset in the formation of the rift that indicated the pathway for magma at depth, and ground water with chemical content that indicated a hot water source. The self-potential survey also pinpointed a definitive bull's-eye on the site (see fig. 3.9); this indicated hot water trapped beneath the surface. Geochemical and geoelectrical surveys also tended to confirm the self-potential survey. The committee members thus assessed the potential of finding a geothermal source at either site within Area A as very promising. Area B was also considered to be a site with high potential, but did not have geologic characteristics quite so promising. In sum, the site selection committee was positive and optimistic about the chances of uncovering a geothermal reservoir.

In November 1974, the geophysics program team members met to revise their section of the proposal. They first reviewed all geophysical evidence that had been collected to date and then evaluated the selected drill site. In reviewing the data, they examined the geoelectrical surveys, the magnetic surveys, the seismic surveys, the geochemical analyses, and the self-potential survey. After considerable debate, they concluded that the geophysical evidence required more study and therefore they could not support the site selection committee's optimism. Moreover, their interpretation of the data suggested that a drilling program to search for geothermal resources was unwarranted. The surveys, although indicating a high geothermal potential, could also be indicative of phenomena other than hot water or steam at depth. Thus, until the geophysics program completed a careful and comprehensive analysis of all data, they considered a drilling program to be premature.

The program team members also decided that if a drilling program were to be funded, despite their recommendations, they still could not support the location chosen by the site selection committee. In elaborating on this position, the geophysics program coordinator wrote in a summary report:

> The Site Selection Committee for the drilling program, which is quite independent in organization from the geophysics program, met in November to select a drilling site for the renewal proposal. As far as the geophysics program is concerned, no special site can be recommended for geothermal exploration, but a hole could be recommended to check geophysical data. The committee chose a site based on self-potential data and geological formation. The geophysical program agreed to go along as a hole at that site will also have value in checking out gravity and magnetic data.(9)

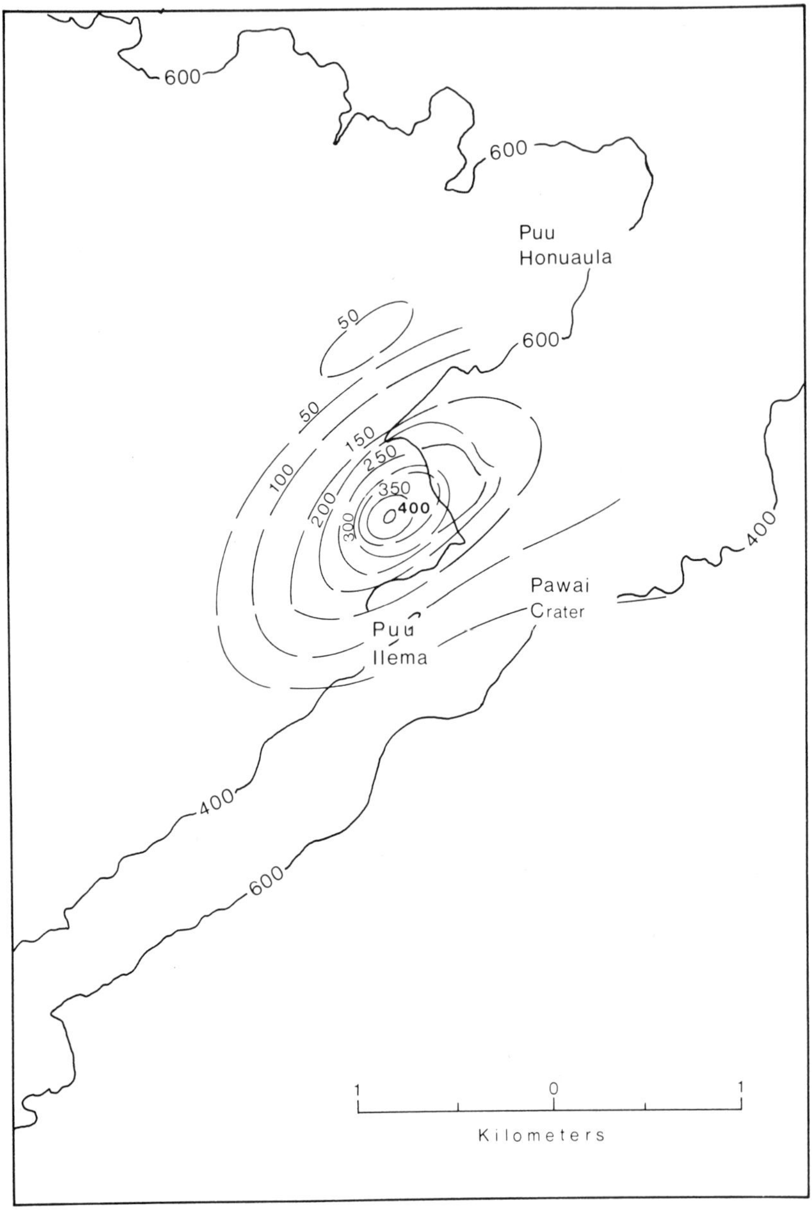

Fig. 3.9. Self-potential mapping of Area A.

Source: Hawaii Geothermal Project, Quarterly Progress, no. 4, March 1, 1974, June 30, 1974, pp. 5-25.

Despite the optimism of the site selection committee and the skepticism of the geophysics program, the revised proposal was written. As could be expected, it reflected the divergent outlooks. On the one hand, the geophysics team emphasized in the proposal that the geophysical data was ambiguous and that further surveys should be undertaken. Moreover, they endorsed a drilling program, not because it had potential for uncovering a geothermal reservoir, but because it could test geophysical interpretations.(10) On the other hand, the site selection committee was very positive about the site and about the potential of finding a reservoir. They stated in the proposal that "probing to depths where we now have no factual information would be most beneficial and we could conceivably arrive in the upper portion of a potential geothermal resource."(11)

The geophysics program coordinator was dismayed by the overall proposal. He later commented that "the fine distinction, exploratory hole to test data vs. exploratory hole for geothermal source, was lost in the writing of the proposal."(12) Nonetheless the revised proposal was submitted to NSF in December 1974.

Apprehension and then approval

While the Hawaii Geothermal Project waited for a decision on funding, the site selection committee received outside opinions concerning the site. Of particular interest was a letter written by George Keller to John Shupe. In the letter Keller expressed the opinion that the committee's reliance primarily on the self-potential survey made him skeptical. In response to this opinion, the committee held a meeting in January 1975 to reassess the site. At the meeting, the committee reexamined their decision in light of all the information which they had received; and they decided that the originally selected site was still the area with greatest potential. The geological structure was favorable, and the geoelectrical and geochemical surveys still indicated subsurface hydrothermal activity.

Later in the month and during February 1975, John Shupe conferred with NSF officials about the revised proposal. He also met with officials from the Energy Research and Development Administration (ERDA). ERDA was the federal government agency responsible for the development of the nation's new energy sources; it would assume from NSF the funding responsibility for an approved drilling program. During the meetings with the officials from both agencies, Shupe received assurance that the proposal was high priority and that it certainly would be approved. NSF even sent a staff worker to Hawaii to advise Shupe and the drilling program team members on preparing invitations to bid for the proposed drilling. Despite the assurances, however, the two agencies had not yet completed the proposal evaluations.

In early March, Shupe received the evaluation; it concluded that there was insufficient geophysical data and analysis to justify a drilling program, especially since the implied intent of the drilling was to tap a geothermal source. The geophysics coordinator commented that he had predicted such an evaluation because there had been insufficient time to conduct a thorough interpretation of the data. The geophysics team thus concentrated on completing a thorough analysis.

By the end of the month, a preliminary interpretation of the geophysical surveys was completed. Using the data from the gravity, magnetic, and microseismic surveys, the geophysics team projected the shape, width, and depth of the dike complex in the proposed drilling areas. The dike complex, or intrusive zone, was projected to begin about 2,953 feet (900 meters) below sea level. It was approximately 1.98 miles (3.2 kilometers) wide, with a vertical extension of about 2.49 miles (4 kilometers) and had the shape of a long horizontal rectangular prism with vertical walls. A complete interpretation of all the data was not made, but early interpretations tended to confirm that there was a possible geothermal resource at Area A.

More important, there was now adequate data and interpretation to justify an exploratory drilling program. The available data and the interpretations were sent to ERDA.

Then, in late April 1975, ERDA informed Shupe that the proposal had been approved. The Hawaii Geothermal Project (HGP) would receive $1,064,000 for the period May 1975-April 1976. This amount, when added to the $500,000 allocated to the project by the state of Hawaii and to the $45,000 given to the project by the Hawaiian Electric Company, amounted to a total of $1,609,000.

Activation

Preliminary activity and key decisions

The project director had actually begun preparing for the drilling activities long before he knew whether HGP would receive drilling funds. As early as October 1974, action was taken to acquire legal permission for the potential drilling. Robert Kamins, the coordinator of the environmental/socioeconomic program, had a University of Hawaii attorney prepare a model right-of-entry permit. The permit was a document granting the landowner's permission for HGP to use his land for drilling. In November, after the site selection committee had chosen the primary and an alternative drill site, Kamins identified the primary site owners as the Tokyu Land Corporation. He sent the corporation's managers a copy of the permit; but after reviewing it, the managers refused to grant HGP drilling permission. Kamins then began negotiations with the owners of the alternative site, the Kapoho Land Corporation. Since this corporation was not developing the site, it agreed to sign the right-of-entry permit. Specifically, the permit granted HGP the rights to enter, to prepare, and to drill the land for a fee of $1. Since HGP was a research project the question of ownership was not relevant. However, if a geothermal resource was discovered, ownership would have to be determined before it could be commercially exploited.

Fully expecting drilling funds to be approved, the project director and the program coordinators continued preparing for the drilling phase. In late January 1975, Abbott, along with the drilling program team and a consultant from the National Science Foundation, prepared invitations to bid on the proposed drilling. The invitations were sent to a number of firms in Hawaii and on the mainland United States. By March, however, drilling funds still were not approved by the Energy Research and Development Administration, so Shupe had to recall the invitations and to wait for the funding decision.

An HGP executive committee meeting was held in late April. At the meeting it was announced, as has been previously discussed, that ERDA was satisfied with the information provided by the geophysics team in March and that firm assurances had been given to Shupe that the drilling would be funded. Official approval of the funds would be forthcoming before the end of the month. The geophysics program coordinator then commented on the selected drill site, Area A. He said that he had completed a more thorough interpretation of the geophysical data, and now he had serious doubts about the geothermal potential of Area A. He felt that Area B, a location which the committee had also considered, had greater geothermal potential. Because of these doubts, another site selection committee meeting was scheduled for May 1975. It was also decided to invite George Keller to the meeting, since he had also expressed doubts about the site.

By May 1975, ERDA had formally approved funds for the drilling; thus the site selection committee meeting took on added significance because it would be the final opportunity to reconsider the drill site. To begin the meeting, the geophysics coordinator commented that Area B was more favorable than Area A because the area of anomalous low resistivity in Area B was considerably larger than in Area A. More disturbing still was that Area A registered high magnetic readings. Since the Hawaiian basalt loses its magnetism above the Curie point, the high magnetism indicated that Area A might not supply enough heat for a geothermal reservoir. Adding support for Area B, George Keller noted that seismic data indicated that rocks in this area had a Poisson's ratio of 0.4. This indicated fractured rock which could possibly allow enough hydrothermal flow to create a reservoir.

In response to these comments, other committee members pointed out that when electrical data from several sources was analyzed, indications were that the anomalous resistivity lows were more definitive in Area A than in Area B. Moreover, the self-potential survey indicated an unambiguous bull's-eye at Area A; and self-potential surveys were found by the U.S. Geological Survey to be the single most useful method for identifying anomalous thermal areas in Kilauea. Finally, there was a moderately high sound intensity of 9 decibels (db) in the vicinity of Area A, which indicated geothermal activity at depth. The high level of magnetism could not be explained, but the magnetic implications conflicted with geochemical and temperature data. A previous geothermal test well located downslope from Area A was 193°F (90°C), and the water in the well contained several times the normal level of silica and chloride, all of which strongly suggested high temperatures at depth. Finally, the Curie point for theolitic basalt could be as high as 572°F (300°C), which was adequate for a geothermal reservoir.

A comparison of the data was then made between the two areas (see table 3.1). It was concluded that the geophysical data, for the most part, was comparable, although magnetic data favored Area B and seismic data favored Area A.

Table 3.1. Comparison of Data Between Area A and Area B

Source of Data	Area A	Area B
Loop-loop inductive soundings	3-5 ohm/m	5-8 ohm/m
Self-potential survey	450 mV	800 mV
Dipole resistivity mapping	8 ohm/m	5 ohm/m
Downslope well temperature	193°F (90°C)	73°F (23°C)
Downslope well chloride content	3,410 mg/liter	–
Downslope well silica content	174 ppm	1 ppm
Ground noise	4-9 db	background only
Magnetism	high	low

The deciding factors would thus have to be geologic, and geology strongly favored Area A. First, Area A lay over the dike complex, while Area B was somewhat astride of it. The dike complex was formed by the consolidation of magma in numerous fissures; thus it formed a potential reservoir of heat. Second, Area A was located directly above an offset in the 1955 volcanic eruptions. These eruptions proceeded to the northeast, stopped, and then resumed in a significantly offset southwest direction. It was believed that a concentration of magma could be located in the vicinity of this offset. Third, Area A was coincident with the epicentral distribution of three separate episodes of shallow earthquake swarms in 1970. These episodes might have been caused by magmatic pressure. Fourth, Area A was located in the vicinities of both seismic activities that preceded the 1960 eruptions and the outbreak of the 1955 eruptions. It was thus believed that Area A was in a zone that had a recent heat source. Finally, there were a number of shallow wells downslope from Area A that were significantly hotter than normal. This supported the hypothesis that ground water flowed downslope through a hot source near Area A.

After all the evidence was again reviewed, unanimity could not be reached and so a vote was taken. All members of the site selection committee, except for the geophysics coordinator, favored Area A. The geological evidence had been the determining factor.

Later in May, Abbott inserted a stake into the ground at the selected drill site within Area A. The overall HGP budget for the drilling phase of the project was then finalized. The allocations were as shown in table 3.2.

Table 3.2. Budget Allocations for HGP Drilling Phase, 1975-1976

Project Director	$ 30,877
Geophysics	237,977
Engineering	155,972
Environmental/Socioeconomic	59,412
Drilling, including:	
Subcontract	979,000
Consulting	40,000
Site Preparation	35,000
Contingency, Testing	50,000
Total	$1,609,151

The consulting firm and the drilling and testing programs

Following the site selection committee meeting, John Shupe met with Abbott and the other drilling program team members and they decided to hire a drilling consultant since they had only limited experience in deep-hole geothermal drilling. The New Zealand firm of Kingston, Reynolds, Thoms, and Alardice (KRTA) had earlier been suggested by an NSF official, and thus Shupe began inquiring about the firm. At the May 1975 United Nations geothermal conference, Shupe learned that KRTA had extensive experience with geothermal drilling in New Zealand, the Philippines, and Central America. The firm was also one of the most respected geothermal consulting firms in the world. It was thus decided to contract KRTA's services and an agreement was worked out with R. Kingston, the firm's managing director.

In June 1975, the coordinator of the drilling program, Agatin Abbott, temporarily withdrew from the project because of ill health, and his colleague, Dr. Gordon Macdonald, a University of Hawaii professor of geology and geophysics, assumed the coordinator's role. Later in June, the invitations to bid for the drilling subcontract were sent to 28 firms. Most of the firms, however, were located on the

United States mainland and it would be too costly to ship a rig to Hawaii. Thus, by July 1, the closing date for bids, only one bid had been submitted. This bid was from Water Resources International (WRI), the Hawaii-based company that had previously drilled Hawaii's only deep geothermal well. It was thus decided to negotiate a contract with WRI.

In July 1975, R. Kingston arrived in Hawaii for consultations with the HGP project team. He discussed the overall program with Shupe and held separate conferences with each of the program coordinators concerning the drilling and testing activities. He also spent a great deal of time discussing the drilling with WRI. He then returned to New Zealand and began preparing the drilling and testing plans.

During July and August, Kingston completed both the testing and drilling plans which were based on the special conditions encountered in Hawaii, the needs of HGP, and the experience of geothermal projects in the rest of the world. The plans were crucial for HGP's success. Not only would they be used as the basis for drawing up contractual obligations between HGP and WRI, but also they would be used to guide day-to-day operations. Consequently, the plans were comprehensive and detailed. The drilling plan was divided into three phases: 1) predrilling site activities, 2) the drilling program, and 3) the site restoration. A brief description of each phase follows.

The predrilling site activities were intended to ensure that the site was adequately prepared and that the contractor had adequately mobilized for a deep hole drill. Specific responsibility for completing each activity was divided between HGP and WRI. Some of the more important responsibilities of HGP included:

- Establishing rights of way and building adequate roads to the site.
- Clearing and grading the drilling area.
- Constructing an 8-foot (2.4-meter) deep drilling cellar of size appropriate to support the drilling rig substructure.
- Implanting in the earth a 30-inch (50.8-centimeter) conductor pipe.

Important responsibilities of WRI included:

- Spreading over the drilling area crushed rock sufficiently fine so as to seal the surface from excessive rainwater percolation.
- Constructing on the site a 180,000-gallon (684,000-liter) water reservoir.
- Providing work offices with supply sheds, fences, fuel, and power.
- Obtaining necessary drilling supplies and equipment, such as liner and casing, cement, valves, hole openers, and various drilling bits.

The drilling phase was planned to be fairly conventional. In order to bore the well, the drilling contractor would use a rotary drilling rig, hole openers, various bits, and additional drill collars. As the drilling proceeded, fluid mud would be injected into the hole to cool and lubricate the bit and to remove the cuttings. It was anticipated that this process could penetrate the most difficult rock formations to a depth of 6,000 feet (1,829 meters).

To encase the well, a series of steel tubes, called casings, would be inserted into the hole. From the surface to a depth of 8 feet (2.4 meters), a 30-inch (762-centimeter) diameter conductor pipe would be placed into and cemented to the sides of the bore. A 20-inch (50.8-centimeter) diameter surface casing would then be inserted into and cemented to the sides of the conductor casing. The surface casing would extend from a depth of 3 feet (0.9 meters) to 400 feet (122 meters), with the length of casing below 8 feet (2.4 meters) cemented to the sides of the hole. Another steel tube would then be inserted into and cemented to the surface casing. Into this steel tube another steel tube would be inserted; and finally, if the drilling struck a geothermal reservoir, a slotted liner extending to a depth of 6,000 feet (1,829 meters) would complete the well. Table 3.3 and figure 3.10 elaborate on the planned well construction.

Table 3.3. HGP Casing Program for Geothermal Well

Casing	Diameter	Depth
Conductor pipe	30 inches (76 centimeters)	0 to 8 feet (0 to 2.4 meters)
Surface casing	20 inches (51 centimeters)	3 to 400 feet (0.9 to 120 meters)
Anchor casing	13 3/8 inches (34 centimeters)	3 to 1,000 feet (0.9 to 304 meters)
Production casing	9 5/8 inches (23 centimeters)	3 to 2,500 feet (0.9 to 762 meters)
Liner	7 5/8 inches (18 centimeters)	2,500 to 6,000 feet (762 to 1,829 meters)

The final phase of the drilling program was site restoration. As planned for, the contractor would have full responsibility for removing all equipment, for disposing of all surplus supplies, and for restoring the site.

The testing program

The testing program, which Kingston completed in August 1975, was based on KRTA's experience in different settings around the world. It was intended to be cost-effective, while producing all the necessary data. As stated in the testing program:

> The testing program which is recommended in this report is based on the experience which has been accumulated in the development of the geothermal fields in New Zealand. A similar

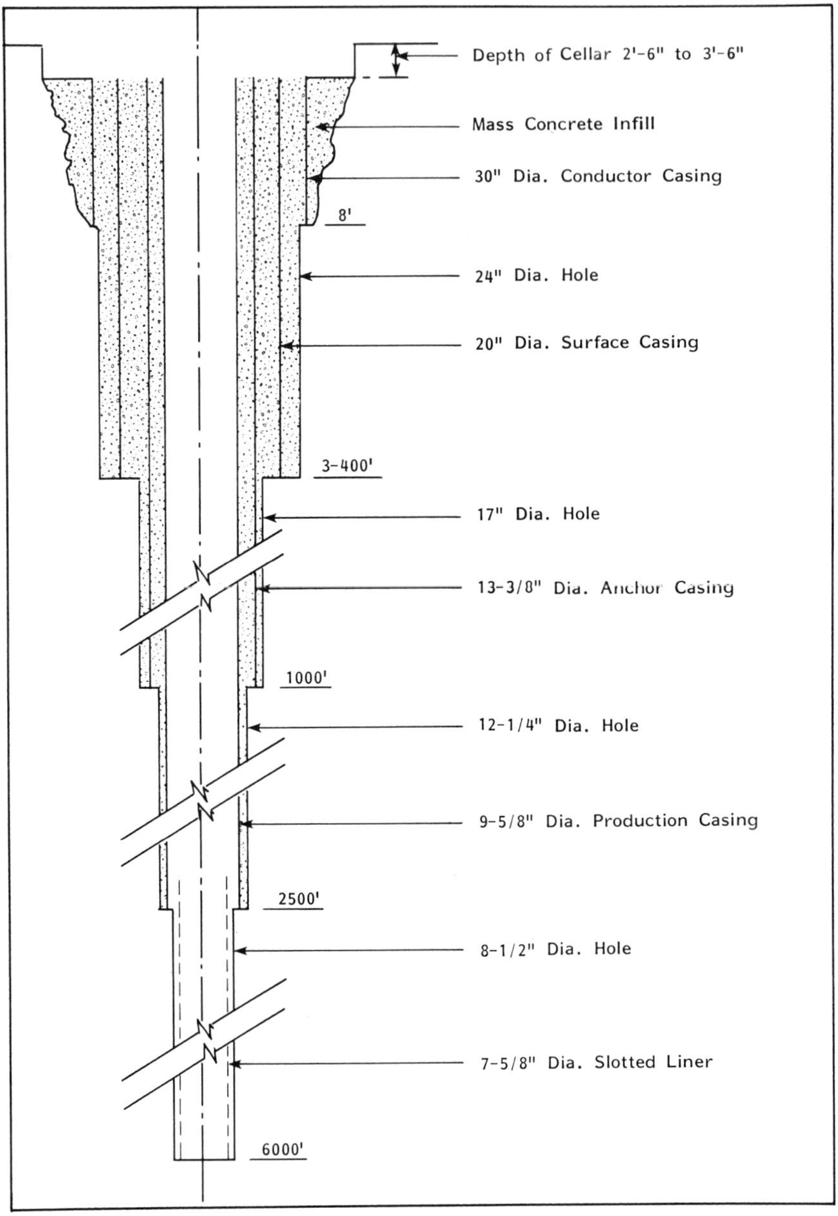

Fig. 3.10. The well casing program.

program is now also being applied in many other countries including the Philippines, Indonesia, Chile, Kenya, Turkey and Nicaragua. The aim of the program is to produce the most useful and factual information which can be obtained from the well, in the most economical manner, and in the minimum time.(13)

The testing program was divided into three stages: 1) drilling tests, 2) drilling completion tests, and 3) output tests.

1. The drilling tests would be conducted during the actual drilling and would help determine how deep to drill and whether there was sufficient geothermal potential to complete the well. Two general types of data would be collected, lithologic data and drilling logs. To derive the area's lithology, geologists would take core samples from the well at approximately 700-feet (213-meters) intervals and cuttings at 5- to 10-feet (1.5-3 meter) intervals. These samples would then be used to complete petrographic and geochemical analyses, which would indicate the structure, composition, and sequence of the formation. The drilling logs, which included neutron, gamma, resistivity, and temperature surveys, would help determine the well's permeability, porosity, and temperatures at various depths. Also a cement bond log would be taken to determine if the cement used for the casing was completely intact. Any flaws or gaps in the cement would have to be corrected with special equipment.

2. Drilling completion tests would be conducted after the well was drilled; they would consist of water loss tests, baseline temperature and pressure measurements, and well starting. The water loss tests were intended to determine the well's permeable zones. To locate these zones, researchers would pump water down the well and take a temperature profile. A gradual change in temperature indicates uniform permeability, while a sudden change indicates a major zone of permeability at or just above the depth of the temperature change.

After locating the possible zones of permeability, the HGP test researchers would try to determine the levels of permeability by pumping water into the well at rates of 100, 200, and 300 gallons (380, 760, and 1,140 liters) per minute. If the rise in water pressure was high it would mean that water in the permeable zones was unable to flow freely out of the well. This would imply low permeability with surrounding rock formation devoid of fissures, fractures, or cracks that would allow a constant flow of geothermal fluid. Lack of permeability meant that the well was nonproductive.

Temperature and pressure of the well would be recorded at regular intervals. When these were stable, they would be the baseline against which all subsequent temperatures and pressures would be compared.

The final aspect of completion testing was to actually start the well. In this process, researchers would remove the top layers of water with compressed air. This would reduce the weight of the column of liquid in the well and enable the pressure from the steam and heat to force the bottom layers of geothermal fluid to flow out of the well naturally and continuously. Once the flow was established, the master valve would be closed and the flow would be shut off.

3. After pressure and temperature had stabilized following the initial well starting, output tests could begin. During output tests, the well would be allowed to flow continuously for periods varying between 7 and 90 days. During each of these periods, the temperature, pressure, mass-flow rate, and heat-flow rate of the geothermal discharge would be measured. These measurements over time would help indicate the well's electrical power potential and life expectancy. The researchers would also collect water and gas samples in order to monitor the chemical discharge of the well.

Mobilizing the project

When Kingston submitted drilling and testing programs to HGP in late August 1975, the drilling subcontract was still being negotiated. The major problem was that Water Resources International was the only bidder, and thus there could be no competitive evaluation. Additionally, the project was jointly funded by the federal government and the state of Hawaii. The acting agencies, ERDA and the University of Hawaii, could not completely agree on what constituted acceptable criteria for evaluating a single source bid. Thus negotiations dragged on.

While the negotiations continued, John Shupe, HGP project director, decided to mobilize the project. He released available funds from the project budget to grade and compact the drill site, to construct the drill rig foundation, and to begin other site preparations. Water Resources International similarly began mobilizing. They purchased a new drill rig and began moving it to the project site. By committing large sums prior to the conclusion of negotiations, Shupe and WRI assumed substantial risk. If negotiations failed, WRI could go bankrupt and HGP could face litigation. Both, however, were confident that points of contention would be resolved satisfactorily. Furthermore, the project was behind schedule, and if they did not begin mobilizing in September, the drilling could not begin until February 1976. It was thus decided to take the calculated risk.

Through September, contract negotiations dragged on. WRI continued to provide the information requested by ERDA and the University of Hawaii; Shupe attempted to mediate and clarify the situation. But federal regulations concerning audit and review processes prevented a swift award of the contract to WRI. Later in the month, R. Kingston, the manager of KRTA, made another trip to Hawaii to help resolve the difficulties. Kingston discussed the subject with the auditors and discovered that they needed more data concerning WRI's cost accounting and the time estimates to complete the drilling. The drilling time was particularly crucial because WRI charged – as was standard practice – by the drilling time expended rather than the depth of penetration. The estimated drilling time was based on WRI's previous experience in drilling Hawaii's only deep geothermal well. For that well, drilling had averaged 100 feet (30 meters) per day. Despite the fact that the HGP well required a hole of considerably greater diameter, WRI assumed it could maintain that rate because it had purchased an improved drill rig that would increase efficiency.

The federal auditors, however, still could not approve the contract because they had difficulty in auditing WRI's records upon which all costs were based. Finally, after correspondence between Shupe and ERDA, the ERDA manager of geothermal development made a special trip to Hawaii to approve the contract. The contract was formally awarded to WRI in November 1975. A summary of the contract's financial accounting appears in table 3.4.

Table 3.4. Subcontract for Drilling Operations

	Contract Estimate
Mobilization	$120,192
Water reservoir	48,077
Casing (to 3,500 feet or 1,067 meters)	100,000
Consumable materials (to 6,000 feet or 1,829 meters) – bits, mud, cement, etc.	202,431
Well testing equipment and services	50,000
Well drilling, 0 to 3,500 feet (1,067 meters)	199,185
Well drilling, 3,500 to 6,000 feet (1,067 to 1,829 meters)	149,500
Demobilization	9,615
Contingency	100,000
Total	$979,000

Project organization in December 1975

In December 1975, just prior to beginning the drilling operation, HGP was organized much as it had been February 1974. John Shupe was the overall project director and director of the management program. Each of HGP's other programs continued to be guided by the original coordinators: Augustine Furumoto directed the geophysics program; Paul Yuen directed the engineering program; Robert Kamins directed the environmental/socioeconomic program; and Gordon Macdonald directed the drilling program. Each program coordinator retained fiscal and operational autonomy for his program, and Shupe provided the overall coordination, direction, and leadership. Additionally, each of the program coordinators along with the project director shared responsibility as co-principal investigators of the grant.

Administrative responsibility for each of the programs was held by each of the coordinators. But since the drilling was subcontracted, authority for the drilling program was fragmented. Macdonald, as coordinator of the drilling program, was responsible for all scientific decisions, such as what tests to conduct and when to conduct them. The drilling consulting firm, KRTA, was responsible for recommending specific technical decisions during the drilling and KRTA provided one employee, Warwick Tracey, as the on-the-job drilling supervisor. The drilling contractor, Water Resources International, was responsible for the daily operational activities of drilling, such as supervising the drilling crew, overseeing the equipment changes, and implementing the operational decisions. E. Craddick, president of WRI, would be at the drill site to oversee these duties.

Some of their responsibilities overlapped. For example, any scientific decision, such as when to take a core sample, impinged upon the drilling supervisor's ability to make technical drilling decisions. Sometimes it might be impractical to stop drilling in order to take a core sample. Thus a technical decision would have to be made that might conflict with a scientific decision. Furthermore, KRTA's drilling supervisor could make recommendations to the contractor about the drilling process, but the contractor was directly responsible for the drilling crews and the operations. KRTA recognized these potential overlapping jurisdictions and, to some extent, tried to clarify them. In the testing program KRTA had stated:

> The University of Hawaii will appoint to the project a Drilling Manager and a Geologist who will supervise the drilling of the well through the Contractor. They will offer guidance and assistance to the Contractor, but in no way will this relieve the Contractor of the responsibility for the drilling operations. The Contractor is required to provide at all times supervision by a competent toolpusher, and skilled crews experienced in the operations of a drilling rig of this scale.(14)

Even with this clarification, the overlapping authority was bound to create some confusion. Thus, during the actual drilling, the HGP project director referred and resolved any differences among the drilling program team representative, the drilling supervisor, and the contractor. Figure 3.11 illustrates HGP's organizational structure as well as its major funding sources in 1975.

Another administrative difficulty of the drilling was that some of the key people would not be at the drill site. The project director had a dual appointment as the University of Hawaii's dean of the School of Engineering; he would have to remain on the island of Oahu. Similarly, the coordinator of the drilling program would not be at the site after January 1976. He would have to return to his teaching duties at the University of Hawaii campus on Oahu. However, plans were worked out to alleviate this situation. Either the project director or the drilling program coordinator would visit the site once a week and they would

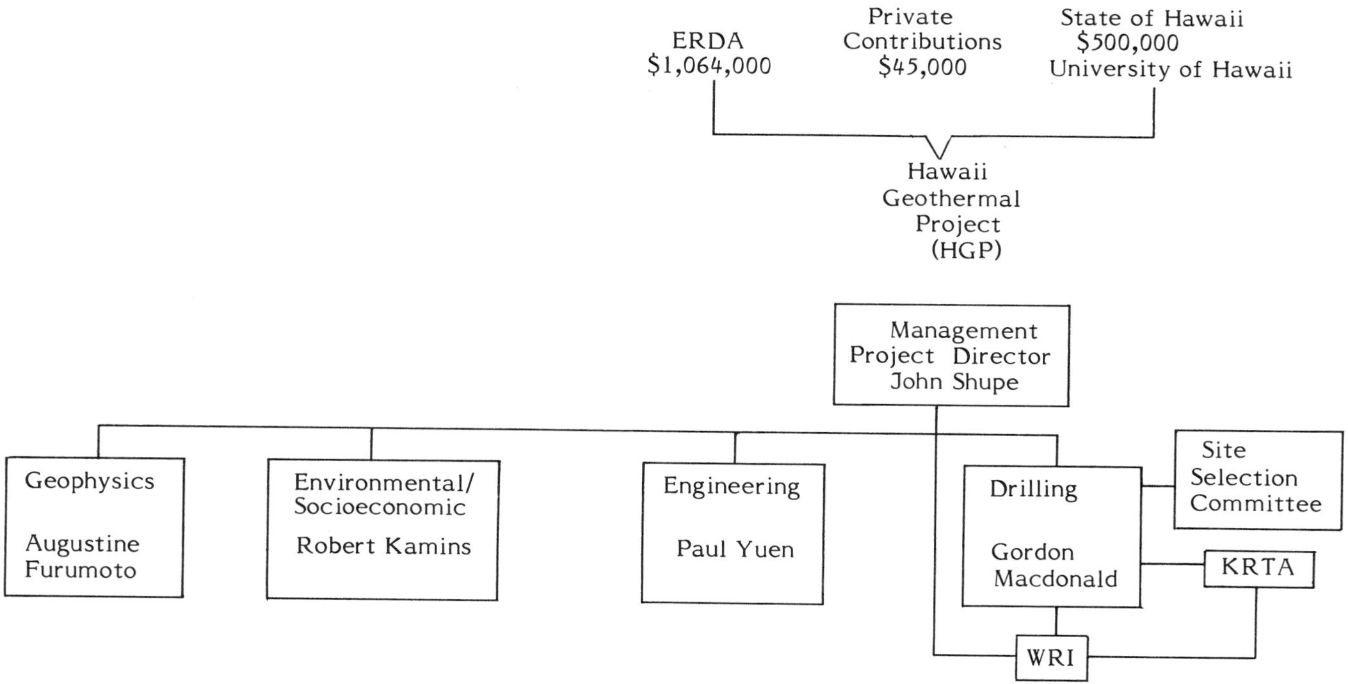

Fig. 3.11. Funding for and organization of HGP, 1975.

communicate with onsite personnel daily. Finally, other project staff would be at the site. D. Palmiter, a graduate assistant with Macdonald, would represent the drilling program and HGP project members from the University of Hawaii campus on the island of Hawaii would help with the drilling. One person in particular, Bill Chen, a professor of engineering at the University of Hawaii Hilo campus, was a key member of HGP's engineering program; he would help with the engineering aspects of the drilling.

PHASE 3: OPERATION, CONTROL, AND HANDOVER

The Drilling Operation

The well site was dedicated on November 22, 1975, and final preparations were made for the drilling operation. WRI assembled the rig, completed installing the plumbing, wired the electrical fixtures, lined the water reservoir, filled it with 180,000 gallons of water, finished constructing the mud pit, and transported to the site equipment such as drill bits, pipes, and casing materials.

Initial drilling

Drilling commenced on December 10, 1975, but progress was slow because the 1955 lava flows had formed a hard basalt layer over the drill site. Drilling the first 400 feet (122 meters) was particularly time consuming. Only limited weight could be placed on the bit as it would not remain vertical. Further delays were caused by the high standards of the casing program, which stipulated that a 20-inch (51-centimeter) diameter surface casing be installed from the surface to a depth of 400 feet (122 meters). This meant that WRI had to initially drill into the lava with a 9-inch (23-centimeter) bit and then use hole openers to progressively enlarge the bore to 15, 20, and 26 inches (38, 51, and 66 centimeters). In early January, the project director realized that this laborious process put the drilling far behind schedule. He therefore conferred with the drilling supervisor from KRTA and then called Kingston, inquiring about the advisability of abandoning the 20-inch (51-centimeter) surface casing and proceeding directly to install and cement in the 13-inch (33-centimeter) anchor casing. Kingston consulted with his staff and with geothermal experts from the New Zealand government's Ministry of Works. The consensus was that if a productive geothermal resource were discovered, the 20-inch (51-centimeter) surface casing would be necessary in order to prevent blowouts. Kingston thus strongly recommended that HGP adhere to the original casing program.

The project director thereupon decided to continue with the original plan. The hole was drilled to a depth of 400 feet (122 meters), and then progressively enlarged with 15-, 20-, and 26-inch (38-, 51-, and 66-centimeter) hole openers. Finally, the surface casing was inserted into the hole and cemented in. This time-consuming process took until early February; it had taken nearly two months to complete the first 400 feet (122 meters) of the well.

Although it proceeded much more quickly in March, the drilling was still considerably behind schedule, as numerous operational difficulties were encountered. At one point, six bolts on the rig jack were sheared off and the rig had to be closed down; at another time, the chain drive linking two drawwork engines broke and had to be repaired; at still another time, the pump and generator engines had to be overhauled because of continuous usage; and, as a routine matter, the drill bits deteriorated rapidly in the dense lava. Additionally, numerous core samples and cuttings had to be taken. Although the coring was necessary and planned for, it was still time consuming and added to the pressures of the drilling schedule.

From an administrative perspective, the drilling problems were handled within the overall framework of HGP's drilling supervision, which, as previously noted, was fragmented. For purely scientific decisions, such as when to core, Macdonald, the director of HGP's drilling program, made all decisions. After Macdonald returned to his teaching duties in January 1976, the scientific decisions were made by his on-the-job assistant, D. Palmiter, who called Macdonald daily concerning the scientific investigations.

For simple operational problems, E. Craddick, WRI president and the onsite drilling contractor, made the decisions. In complex and difficult drilling situations, Craddick consulted with Warwick Tracey, who was both the on-the-job KRTA consultant and the drilling supervisor. However, since the duties, responsibilities, and overall authority of drilling supervisor and contractor were not absolutely defined, some conflicts arose when there was a difference of opinion. In one situation, for example, WRI's 20-inch (51-centimeter) hole opener had to be inserted into the 20-inch (51-centimeter) surface casing in order to continue enlarging the hole for the anchor casing. The hole opener would not fit, so the drilling supervisor recommended that the contractor undercut the periphery of the cutters and then rebuild them back up again with hard facing electrodes. The contractor, however, decided to simply cut off the periphery and remove the hard facing of the cutters.

Whenever there was a difference of opinion on a critical activity, the drilling supervisor would have to call Shupe and Kingston. For example, in February and March 1976, the drilling supervisor recommended that the contractor clean up the site and make adequate provisions for the disposal of reject mud and cuttings from the drilling. No site cleanup, however, was undertaken. This forced the drilling supervisor to write to Shupe and Kingston and comment: "Disposal of reject mud and cuttings is still a major problem, and as a result, the site is a mess. The contractor does not seem to appreciate the magnitude of the problem, and his disposal gear is primitive, if not inadequate."(15) Shupe and Kingston then conferred with the contractor and he agreed to improve the disposal of reject mud and cuttings. If the problem was serious enough, an onsite conference would be held. In resolving the casing program, for example, Shupe contacted Kingston and arranged for him to fly in from New Zealand to meet at the drill site. Prior to the conference, Shupe conferred with the other HGP program coordinators.

Two significant problems

During February and March 1976, two significant problems arose, which went beyond the scope of the operational drilling management. First, on March 18, HGP project staff projected that an additional $257,000 was required to complete drilling the well to the planned goal of 6,400 feet (1,951 meters). The deficit was caused because the process of installing the surface casing had taken more drilling time than originally estimated. Since payment to WRI was based on actual drilling time, additional funds were required.

The project director first contacted ERDA, and after clarifying the situation through a letter and several phone conversations, received an additional $150,000. ERDA was extremely supportive because the well was so near completion. WRI was also anxious to see the well be successful, and agreed to donate $60,000 of its time to finish the well to 6,400 feet (1,951 meters). Finally, the project director met with the program coordinators. He had previously informed them of the drilling deficit and had discussed the possibility of reallocating funds. Now, he stressed that the probability of encountering a productive geothermal resource would be increased if the well were drilled to its original target depth. Moreover, ERDA, the funding agency, was extremely interested in completing the drill to 6,400 feet (1,951 meters). The program coordinator thus agreed to shifting $47,000 from the research programs to the drilling.

The second significant problem centered on the internal differences of opinion within HGP. As discussed earlier, there was a lack of unanimity on many of the HGP policy decisions. There had never been, for example, unanimous agreement on the decision to establish a drilling program or where to locate the drill site. Now during the drilling, there was still some difference of opinion about the potential of the site. The geophysics program coordinator, after examining more data, believed that the drilling should be terminated at about 4,000 feet (1,200 meters). Among his reasons were: 1) if a geothermal reservoir existed, it should be located between the water table and the dike complex, and 2) gravity, magnetic, and deformation data now indicated that the top of the dike complex was situated at depths between 1,640 feet (500 meters) and 4,000 feet (1,220 meters).(16)

The drilling program coordinator had a different opinion. He believed that drilling only to 4,000 feet (1,220 meters) was meaningless for the following three reasons: 1) the purpose of drilling was to discover whether rocks or structures favorable to a geothermal reservoir existed at depths of 6,000 feet (1,829 meters), 2) there were many dikes in the rift zone; they could be located at any depth below the dike complex, and 3) there was a possibility of a geothermal reservoir in the interdike compartments below 4,000 feet (1,220 meters).(17)

The issue was resolved in a series of meetings and discussions among the project director and the program coordinators. It was decided that since there was enough money for only one deep drill, it would be counterproductive to stop just when the most difficult portions of the well had been completed.

Completing the drilling

In early April 1976, after the production casing was installed to a depth of 2,200 feet (671 meters), the drilling proceeded rapidly. By April 27, the well was drilled to its target depth of 6,400 feet (1,951 meters). The final 4,200 feet (1,200 meters) of the well had been bored in less than three weeks, with no difficulties or significant problems. Since the drilling mud at 6,000 feet (1,829 meters) was about 145°F (63°C) and heating up as time passed, it was known that the well was hot. However, it was not known how hot the well was or whether it was productive.

The first measurements, the well logs, were conducted in late April, after the well had been drilled to its target depth. WRI used its own Gearhart-Owen equipment and hired an operator to conduct neutron, gamma, self-potential, and resistivity logs; these logs would provide indications of the formation's permeability and porosity. Temperature logs were also taken; they indicated that at a depth of 4,000 feet (1,220 meters), the well temperature exceeded 300°F (135°C), the upper limit of the equipment. Thus, precise temperatures were unknown, but it was certain that the temperatures were favorable.

The contractor then conducted a cement-bond log. This log would determine the integrity of the concrete used to cement the casing. Any flaws in the concrete would have to be corrected before completing the well. The cement-bond log indicated a lack of bond, and by implication, gaps in the cementing, from 40 to 220 feet (12 to 67 meters) and from 320 to 868 feet (97 to 264 meters).

Following the well logging, an HGP staff meeting was held and it was determined that three tasks would have to be undertaken. First, the gaps in the cementing of the casing would have to be filled or the well's steam pressure and thermal stresses could severely damage the casing. The gaps would have to be filled before proceeding further. Second, a slotted liner – a 7 5/8-inch (18-centimeter) diameter steel tube with 8 slots per foot (0.3 meters) – would have to be run into the well from a depth of 2,200 feet (670 meters) to a depth of 6,400 feet (1,951 meters). The slotted liner was essential. It would prevent the well's bottom uncased section from caving in, and its 2-inch-by-3/4-inch (5-centimeter-by-1.9-centimeter) slots would allow steam or hot water to enter the tube from the side and be transported to the surface. Third, the drilling completion tests would have to be conducted.

These three tasks would require funds above and beyond HGP's budget, since all funds had already been used to complete the drilling.

It was estimated that the three tasks would require an additional \$248,000. The project director therefore again contacted ERDA officials, and they invited him to present a progress report of the drilling at a meeting of the ERDA geothermal coordinating group in late April. At the meeting the progress of the drilling was reviewed and requests were made for additional funds to complete the well and to conduct the well testing. Additional funds were then promised to the project and the director returned to Hawaii and ordered 4,500 feet (1,373 meters) of slotted liner. Arrangements were also made to remedy the gaps in the casing.

Later in the month ERDA officials formally notified HGP that they were releasing \$85,000 immediately and would release an additional \$175,000 as soon as it could be transferred from the central office in Washington. They also mentioned that they would provide an additional \$300,000 in 1977 to comprehensively test the well.

Completing the well

After the money was secured in early May 1976, Gearhart-Owen perforating equipment and personnel were contracted to fill the gaps in the casing. Also purchased were the valves, gauges, and drilling rig time necessary to install the slotted liner and complete the wellhead plumbing. Finally, the project director arranged for KRTA to continue as consultants and he released funds to purchase testing equipment such as separators and sampling bottles.

The special equipment and operators necessary to fill the voids in the cementing were not available until late May, so work on the well was halted for three weeks. Then, in late May, the special personnel and equipment arrived and the correcting of the casing gaps began. In a four-day period, the special operators perforated the cement with controlled explosive charges, and then forced cement through the perforations into the cementing gaps. The contractor then ran a cement-bond log and determined that the gaps were filled.

In early June, the slotted liner was inserted into the well without any problems. The contractor first used water to cool the hot mud, which had hardened in the bottom of the well, and then obtained circulation. Next the bottom sections of the well were reamed out and the slotted liner was run into the well from 2,200 to 6,400 feet (671 to 1,951 meters). After the liner was installed, the mud in the liner was flushed out, and the well head plumbing including the side and master valves were installed.

While WRI was completing the well in June, the HGP project staff, aided by KRTA, conducted the drilling completion tests. HGP project staff, directed by Bill Chen, a professor of engineering at the University of Hawaii at Hilo, first did a survey of the well's baseline temperatures. Temperature at 4,000 feet (1,219 meters) was about 338°F (170°C); between 4,400 to 6,000 feet (1,341 to 1,829 meters) it remained constant, ranging between 464° and 482°F (240° and 250°C); and below 6,000 feet (1,829 meters), the temperature exceeded 500°F

(260°C). Subsequent temperature surveys, conducted before and after the water pumpdown tests, indicated that the well was continually getting hotter as more and more of the drilling and mud cuttings were cleared from the liner. On June 15, eight days after the second pumpdown tests, the temperature at 4,000 feet (1,219 meters) was nearly 572°F (300°C); and at 6,400 feet (1,951 meters) the temperature exceeded 608°F (320°C).

The other significant completion test was the water-loss or pumpdown test, indicating the level of permeability. To conduct this test, HGP researchers pumped water into the well at rates of 0, 100 (380), 200 (760), and 300 gallons (1,140 liters) per minute, and simultaneously measured the back pressure with a Kuster pressure gauge. If the pressure increased over 150 psi, it would indicate a very poor permeability and nonproducing well.

The initial measurements indicated that the rise in pressure when pumping water into the well between 0 to 300 gallons (1,140 liters) per minute exceeded 700 psi. More pumpdown tests were conducted because the high back pressure could have been created by drilling mud or cuttings obstructing the slotted liner and preventing the inflow of geothermal fluid. However, subsequent tests also indicated high rises in pressure and relatively impermeable conditions (see table 3.5).

Following these tests, WRI demobilized its part of HGP. Employees cleaned the site, removed excess material, removed drilling equipment, and dismantled the rig. However, the most dramatic aspect of the drilling completion test still remained to be done – starting the well.

On June 22, members of HGP's engineering program, who would conduct the output tests, attempted to start the well by airlifting. A hose attached to an air compressor was inserted into the well and the compressor was started. The compressed air then evacuated upper layers of cold liquid to lighten the well's liquid column and to allow it to be heated by the hot geothermal fluid and steam at the bottom. The first attempts to start the well were unsuccessful. Finally on July 2, 1976, the well was flashed and allowed to flow for five minutes.

With the initial flashing of the well, the drilling completion tests were concluded. However, HGP would still have to conduct the formal well tests to determine the well's potential and productivity.

The Testing Period

The testing period of the HGP-A well lasted from July 1976 to June 1978 and was funded by the Energy Research and Development Administration (ERDA), which in 1977 was consolidated into the national Department of Energy (DOE), and the state of Hawaii.(18) The total funds awarded to HGP amounted to $439,000 during this time period.

Organization of the Hawaii Geothermal Project in July 1976

The top priority during this period was, of course, the well testing and each of the programs was redirected to reflect this priority. The

Table 3.5. Summary of Pumpdown Test

Date	GPM*	Time of Flow (minutes)	Volume (gal)	Back Pressure (psig)
June 6	340	46	15,640	700
June 6	108	105	11,340	500
June 6	108	60	6,480	500
June 6	200	55	11,000	600
June 6	300	70	21,000	700
June 6	530	10	5,300	750
June 6	630	7	4,410	800
June 6	300	8	2,400	700
June 6	200	5	1,000	600
June 6	100	6	600	500
June 7	300	3	900	–
June 7	100	180	18,000	300
		Total	98,070 gal	

*Gallons per minute.
Source: Hawaii Geothermal Project, HGP-A Reservoir Engineering, September, 1978, p. 7.

project director, although still responsible for the overall management, coordination and leadership of HGP, now increasingly concentrated on policy, planning, and strategy for future geothermal development in Hawaii. The HGP executive committee would also continue to play a large role in policy making for the future.

The engineering program, assigned the responsibility of conducting the well tests, became the most visible, as well as the most heavily funded, program. Consistent with previous policy, Paul Yuen, the engineering program director, had responsibility for allocating funds and making substantive program decisions. Bill Chen would be at the site to supervise the tests.

The geophysics program was consolidated into a geoscience program. Researchers of the new program would not only synthesize the geological, geophysical, and geochemical surveys into an integrated interpretation, but also would oversee all new scientific inquiries, such as environmental monitoring and the measuring of physical samples from the well tests. The change in program designation thus reflected the program's broader and more encompassing scope. Named as co-ordinator of the new program was Charles Helsley. Helsley was a professor of geology and geophysics and was also the newly recruited director of the Hawaii Institute of Geophysics. Because he had worked previously with other deep drills, it was believed that he had the knowledge and experience necessary to synthesize all the geological and geophysical data into an integrated interpretation. This was particularly important since there had been many conflicting interpretations and none was completely consistent with the actual findings from the drilling.

The environmental/socioeconomic program was nearing completion of its research activities, and would be phased out by Robert Kamins, the director, after two crucial reports were finished. These reports were: 1) an environmental baseline study of the Puna area, and 2) an assessment of geothermal development in Puna. Kamins expected both reports to be completed by January 1977.

Initiating the output well tests

In July 1976, after the well had stabilized, the engineering program researchers prepared to conduct the output tests to determine the well's potential productivity and life span. As described by KRTA in the testing program, the well would be allowed to discharge continuously for extended periods, during which the temperature, volume, pressure, and chemical content of the discharge would be measured. After the discharge period, the well would be shut down, and over the next few weeks researchers would record the well's temperature and pressure at various depths. If well temperature and pressure recovered rapidly, it would indicate a potentially large reservoir.

The first extended discharge test was scheduled for July 22. Since a good deal of public enthusiasm and attention was now focused on the project, it was planned to be a four-hour public display. In this regard, HGP had been primarily a research project with no real tangible product to capture the imagination. But now, there was more than an exciting idea: the Hawaii Geothermal Project could actually display, for the first time in Hawaii, human-controlled geothermal energy. The public display was dramatic and impressive. Flowing continuously for four hours, the well sent geysers of steam and liquid over a hundred feet into the air. Because no muffling or silencer was installed for the first test, the noise exceeded 122 decibels – the sound of a 747 at takeoff.

The test results themselves were very promising and even exceeded the expectations of the KRTA consultants. Lip temperature exceeded 302°F (150°C) for the entire four hours while lip pressure from the 6-inch (15-centimeter) discharge pipe was 23 psig at the end of four hours. Subsequent temperature surveys indicated that the well recovered in about seven days and that the major production zones occurred between 3,500 (1,067) and 4,500 feet (1,372 meters) and at 6,400 feet (1,951 meters).

Based on the test results, the engineering program researchers were able to make some tentative conclusions. First, based on the data, they deduced that the well had a mass flow rate of 166,000 pounds per hour – assuming an enthalpy of 800 Btu/lb and 15 percent efficiency, the well could generate 5 MW of electricity. They also noted that fluid and steam were flowing naturally into the slotted liner, indicating that there was greater permeability than they had assumed from the water-loss tests. Longer tests, however, would have to be conducted before any conclusions could be made about the well's productivity.

Before conducting such tests, however, the engineering researchers would have to construct more elaborate testing equipment. Thus, over the summer of 1976, the researchers added to the well a silencer-separator to muffle the loud noise of the discharge and to separate the water from the steam. Also added to the well were a twin cyclone sampler, which would obtain gas and water samples for chemical analysis, and a calorimeter to measure specific enthalpy. Figure 3.12 illustrates these instruments on the well.

Extended flow tests: problems and progress

In November 1977, with the new instruments and equipment installed, HGP researchers conducted a two-week test flow. The tests were extremely encouraging and demonstrated that the well was capable of discharging continuously for a two-week period. Most prominently, well output stabilized after25 hours. HGP researchers collected other data, such as noise level, temperature, steam quality, and water and gas samples; and they calculated the enthalpy of the discharge and its thermal power. (See "Phase 4: Evaluation and Refinement" for complete data.) By the end of 25 hours of continuous discharge, thermal power had stabilized to about 22 MW and at the end of the test it was still at 20 MW. Thus, from the perspective of demonstrating the well's generating power, the November tests were a success.

However, since the tests were, for the first time, being conducted under real world conditions, they also introduced a new issue – the well's impact and interrelationship with the human community in Puna. This issue was underscored by the land use near the well. North and south of the well were undeveloped areas covered by recent lava flows and sparse vegetation. But to the east and west there were homes. Twelve families lived within one mile of the well and within two miles there were several residential tracts that were being developed. The largest of the residential tracts was Leilani Estates, with a total of 2,146 house lots.(19) Although, at that time, only 50 families were living in Leilani Estates, they had formed an active community association that represented 2,036 of the owners of the lots.(20) Two other residential tracts nearby included Lanipuna Gardens and Nanawele

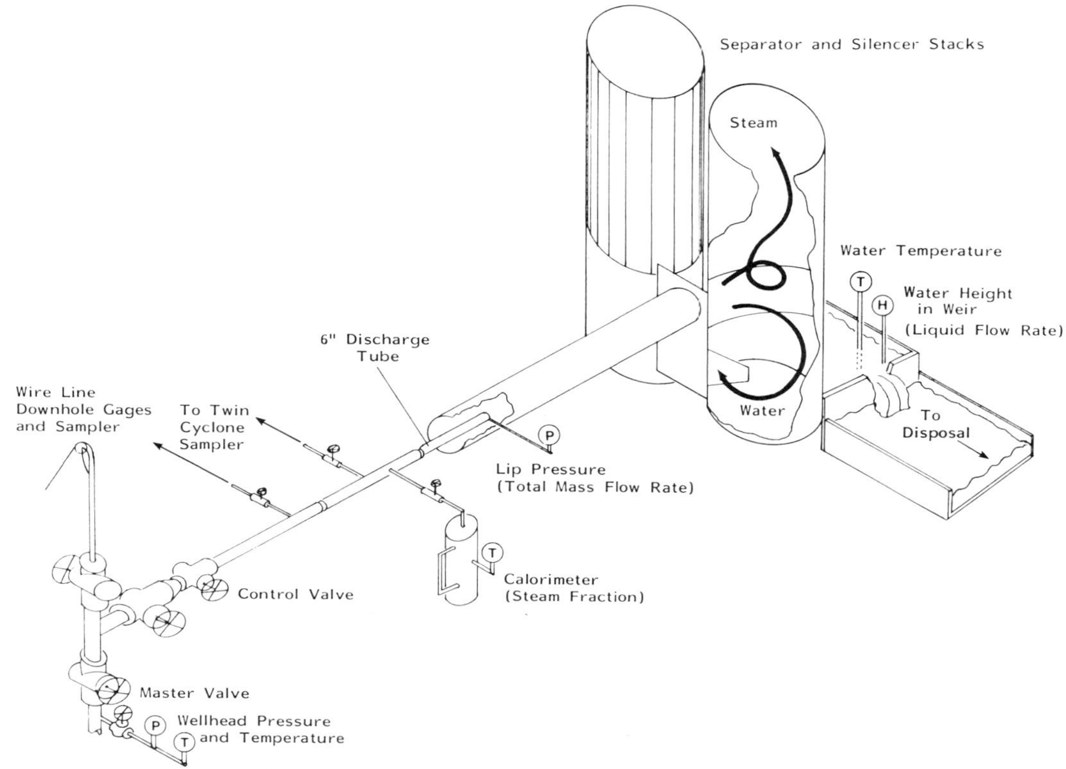

Fig. 3.12. HGP-A well-head testing equipment.

Estates; both had few actual residents but both were being planned for development. Within the immediate vicinity of the well there is also a state park and one paved road, with usually very light traffic (see fig. 3.13). A statement from the environmental baseline study gives one an idea of the setting:

> In most hours of most days, the quiet of the roads in this portion of the Puna District is not much disturbed by passenger cars, or by an occasional truck or bus, most of them traversing the distance to beaches on the Puna Coast. (The well drilling itself generated a fair amount of traffic, not only from the dozen members of the drill crew and scientist observers, but also from some tour buses, whose operators were glad to find a drilling rig to add to the attractions of tourism in this outstandingly quiet corner of the Island of Hawaii).(21)

Further from the well site, but within a several mile radius, are the communities of Pahoa, Kapoho, Opihikao, Kalapana, and Kaipu. Many of the residents of these communities are native Hawaiians, who have lived in Puna for generations and who have adopted a rural agricultural life-style. The entire district of Puna, in fact, is predominantly rural and agricultural. There are extensive cultivated fields of sugar cane and papaya, with smaller areas utilized for growing guavas, oranges, and macadamia nuts. Additionally, there are several small family enterprises growing tropical plants, such as anthuriums and orchids. Interspersed between the villages and the low-density residential tracts are areas of lush tropical vegetation, conservation zones, and several forest reserves (see fig. 3.14). In sum, the Puna district maintains a somewhat traditional rural Hawaiian setting. It is sparsely populated, little developed, primarily agricultural, and outstandingly quiet.

In this setting, the November flow tests began. And, within a few days, the Leilani Community Association objected vigorously to the noise from the well's discharge. The well had been muffled since the July test, which created a noise of 122 decibels, but even with the muffler, the sound at the roadside was 87 decibels, while one mile away the noise was projected to be 70 decibels and at two miles, 40 decibels. The Environmental Protection Agency (EPA) recommended 55 decibels as a tolerable daytime level for residential areas, but the well discharged continuously and within an extremely quiet surrounding. Thus, the community association contacted their city council representative and also contacted state officials, demanding that the tests be halted. One resident stated that the noise was intolerable; he described the sound as a "bloodcurdling banshee howl." Other residents pointed out that the area was outstandingly quiet and that many of them had moved to the area for just that reason. Moreover, their whole way of life was being disrupted by the nuisance of the noise. The county officials responded that they would try to help. But, after investigating the problem, they discovered that there were no noise standards governing residential areas on Hawaii. Thus, the county officials called HGP and asked them to confer with the residents of the area.

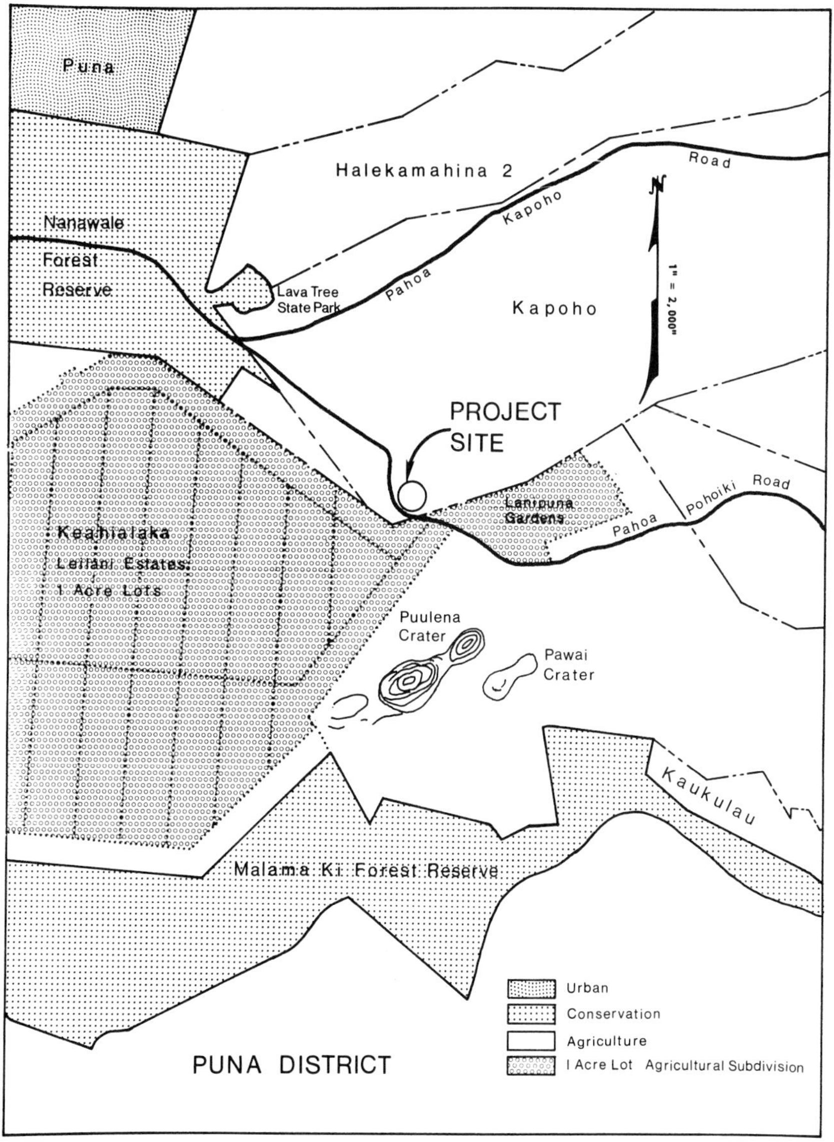

Fig. 3.13. Land use within the immediate vicinity of HGP-A.

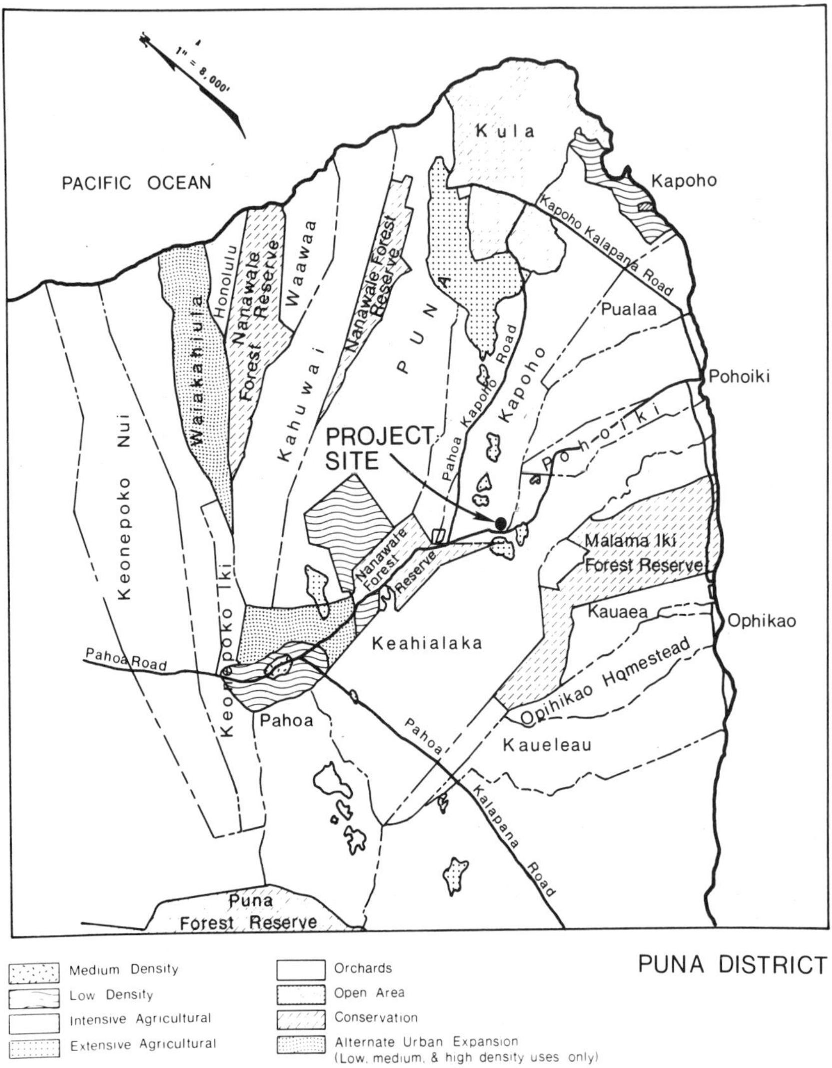

Fig. 3.14. Puna district.

Eventually, the project director met with the residents and agreed to try and improve the muffling on the well. He pointed out, however, that the well was experimental and that the tests would be a nuisance for only a limited time. To further cooperate with the residents, Bill Chen, who lived on the island and was helping oversee the tests, agreed to confer with them before tests and to meet with them when they wanted.

Well tests in December and January 1977

In December, the engineering program researchers conducted a flow test to obtain downhole temperature and pressure measurements while the well was discharging. The tests revealed that downhole temperatures approached 662°F (350°C), one of the highest temperatures ever recorded in a geothermal well. Also encouraging were measurements showing that the well's temperature, pressure, and mass flow rate had increased significantly since the November test.

During the December tests, residents again strenuously objected to the noise. Thus additional muffling was put on the well. This satisfied the residents, and they agreed to a two-week test in January 1977.

The January 1977 tests were intended to determine the well's potential generating capacity using different diameter orifice plates. This would allow HGP researchers to determine which orifice was the optimum size to be used for generating electricity. The results of the test appear in table 3.6.

Well tests, March to May 1977

Following the January tests, Paul Yuen and Bill Chen decided that the well should be tested for a 90-day period to collect enough data to project the well's electrical generating capacity over a 30-year period. The test began on March 21, with the discharge line fully open. A few days later a 3-inch (7.6-centimeter) orifice was placed into the line since this was probably the discharge diameter that would be used to generate electricity.

Soon after the discharges began, residents of nearby Pahoa, Nanawele Estates, and Leilani Estates objected vigorously to the well's odor. The odor was caused by hydrogen sulfide, which was emitted into the air during discharge. In large amounts, it is lethal; in small amounts, it is obnoxious. It has the smell of rotten eggs and the human nose is extremely sensitive to its odor – able to detect it in quantities as small as three parts per billion. The residents thus registered complaints with the Hawaii County Council, the state Department of Health, and directly to the project director. Some residents complained that the hydrogen sulfide fumes were environmentally dangerous and that the odor was a health hazard. Other residents complained of respiratory problems, while a doctor blamed the fumes for causing increased incidences of sinus difficulties, asthma, bronchitis, diarrhea, and dermatitis.

Table 3.6. Throttled Flow Data for January 26-February 10, 1977

Orifice Size (Inches)	Total Mass Flow Rate (klb/hr)*	Steam Flow Rate (klb/hr)	Steam Quality (Percent)**	Wellhead Pressure (psig)	Wellhead Temp. (^{o}F)	Possible Electrical Power Output (Mwe)
8	101	64	64	51	295	3.3
6	99	65	66	54	300	3.4
4	93	57	64	100	338	3.5
3	89	54	60	165	372	3.5
2 ½	84	48	57	237	401	3.3
2	81	43	53	293	419	3.1
1 3/4	76	39	52	375	439	3.0

*klb = 1,000 pounds
**Steam quality = fraction of steam in total flow

Source: Hawaii Geothermal Project, April 1, 1977, p. 19.

A Department of Health representative met with the residents and explained that there were no ambient air quality regulations for hydrogen sulfide; there were, however, federal regulations for industry which specified ten parts per million over an eight-hour day. The HGP well was discharging about three parts per million. The Department of Health official thus explained that the odor was a nuisance that would have to be controlled, but that he could not force HGP to do anything.

After residents protested a while longer, the project director met with the Leilani Community Association. At the meeting, he emphasized that the well was still experimental and that this was probably the last time any extended tests would be conducted before HGP installed scrubbers to virtually eliminate odor. He also noted that natural volcanic eruptions were intensifying the fumes. However, he did acknowledge the nuisance factor and he agreed that if there was any indication of a health hazard he would stop the test immediately. In response the residents cited numerous health problems and indicated that their water supply was largely from a rain catchment system and therefore any fumes in the air were likely to endanger their water. After discussing the problems further, it was agreed to shorten the test period.

The test period was shortened to 42 days and ended on May 9. Although the full 90-day flow test would have provided considerably more information, the engineering program researchers had sufficient data to make some tentative conclusions about the electrical potential of the well. The most promising aspect of the data was that the well output had stabilized and thus extrapolations indicated that the well could continue to generate 3.0 megawatts of electricity over a 30-year period. (See "Phase 4: Evaluation and Refinement" for complete well test results.)

Completion and Handover

Policy and direction of HGP, 1976-1978

While the engineering program team was conducting the well tests during 1976 and 1977, John Shupe and other members of HGP were formulating plans, policies, and strategies for promoting Hawaii's geothermal energy development. In order to begin the planning activities, Shupe, during the middle of 1976, discussed with ERDA officials the possibility of their funding a long-term, large-scale geothermal program in the state of Hawaii. He discovered that ERDA would be reluctant to support any such program for the following reasons:

1. ERDA was doubtful that Hawaii's geothermal energy development had relevance and significance for the nation as a whole. In this respect, HGP had emphasized the importance of obtaining geothermal knowledge of island volcanic regimes, but had never emphasized the potential national spin-offs of such research. If Hawaii was

to continue to receive federal funds, a strong case would have to be made for Hawaii's contribution to the nation's geothermal energy development.

2. Since ERDA was a national federal agency, there was concern that the concentration of support – over $2,000,000 thus far – to one specific geographical location was unbalanced. Requests for support of geothermal projects had been received from over 100 different locations and many would not be funded if Hawaii received a disproportionately large share of the national budget.
3. Finally, ERDA was skeptical about funding development projects through universities because ERDA's objectives were application and utilization. The more practical and effective approach, which was normally ERDA policy, was to fund projects through industry and other "real world" organizations, such as utility companies and energy-related corporations.

Since ERDA funding was essential to future geothermal development efforts, Shupe prepared a number of recommendations that addressed these concerns. For the immediate future, he had two suggestions. First, he recommended that HGP be dissolved at the end of 1977, following the completion of the final research reports. However, an essential corollary to HGP's dissolution would be the formation of a geothermal development consortium, composed of the State of Hawaii, the County of Hawaii, the University of Hawaii, and the Hawaii Electric Light Company (HELCO). Second, he recommended that the consortium plan and implement a coordinated program of geothermal research and development for the state, with its first objective being the construction and operation of a small demonstration plant, powered by the HGP-A well. These two recommendations were intended not only to satisfy ERDA's policy of funding "real world" projects, but also to achieve HGP's ultimate goal of generating geothermal electricity on a practical scale.

These two recommendations required immediate and vigorous action. Thus, in late 1976, Shupe asked Hideto Kono, the state energy resources coordinator, to assume the lead role in organizing and directing a geothermal consortium. Kono agreed and he formally contacted officials from the Hawaiian Electric Company, and from Hawaii County. The state and county governments and the electric company were already cooperating in the development of geothermal power through their active participation in HGP. These groups thus agreed to participate in the consortium, viewing it as a natural outgrowth of the cooperative effort in HGP and as the appropriate organizational entity for developing Hawaii's geothermal power. Also included in the consortium was the University of Hawaii, represented by HGP.

Formally organized in early 1977, the consortium was named the HGP-A- Development Group (HGP-A D/G). Kono was selected as the group's executive director and Shupe was authorized to negotiate for the group in any dealings with ERDA. The members of HGP-A D/G held several meetings during the early part of 1977 and they agreed that their immediate goal was to build a geothermal power facility to demonstrate the feasibility of generating geothermal energy from the HGP-A well. To achieve this goal, the group first prepared a funding proposal for ERDA. Each consortium group contributed its special expertise and the state government provided the funds to contract TRW, Inc., a geothermal consulting firm, to aid in preparing the proposal. Completed in April 1977, the proposal was submitted to ERDA.

HGP-A D/G proposal to ERDA: general design of experimental station

The proposal requested funds to build an experimental geothermal facility at the site of the HGP well. The facility was envisioned to include three basic components: 1) the power generation system, 2) the experimental system, and 3) the support system.

The first component, the power generation system, was to be composed of a turbine generator, condensor, cooling power, antipollution devices, electrical conversion and distribution apparatus, and either percolation ponds or a reinjection well. It was envisioned that the steam and the hot fluid from the well would be piped directly to a separator where they would be separated. The hot fluid would then be piped to the reinjection well or to percolation ponds for disposal, while the steam would be routed to a demister. The demister would remove any remaining moisture in the steam and then allow the dry steam to exit into the turbine at 52 klb/hr at 160 psig. This would produce 3 MW of electricity.

The electricity would go to a specially constructed substation and then be fed into HELCO's power grid. However, since the grid at the substation could accept only 2 MW of electricity, the 1 MW surplus would either be fed into a load bank where it would be dissipated, or it would be used to supply the facility's electrical needs. Power plant operations were to be handled in a motor control and instrumentation center where the turbine regulator, the voltage regulator, voltmeters, ammeters, pressure meters, and other process control instruments would be contained.

The plant was intended to operate in an environmentally sound manner. The steam used to run the turbine was to be exhausted into a condensor, where cool water would condense the steam, leaving water for noncondensible gases. The gas would flow to a cooling tower and, after being cooled, it would either be piped to the reinjection well or recirculated to the condensor. The noncondensible gases, primarily hydrogen sulfide, would be treated in a pollution abatement system and, when safe, released into the atmosphere.

The second component of the facility was to be the experimental system, which was to consist of three test pads, one to conduct electrical geothermal experiments and two to conduct nonelectrical experiments. It was envisioned that nonelectrical experiments could include testing the environmental effects of geothermal fluid, developing a heat exchanger, and developing methods for sampling fluids and gases. The electrical experiments could include testing small geothermal generators, developing a total flow turbine, and evaluating corrosion and scaling problems.

The third component of the facility, the support system, was to include the supply buildings, the administration offices, the repair and maintenance areas, the buildings for the power station, the electrical substation, the instrumentation and equipment needed to monitor the facility, the electrical lighting, and all access and service roads. A detailed layout of all three components would later be integrated into a final design.

Management plan

The proposal called for three stages of activity. The first stage was to be the overall planning; specific activities were to include devising a system of project management, drawing up a preliminary budget and a work schedule, and establishing a monitoring program to control activities. Also during this period, an environmental impact statement would be completed and a design contractor would be selected to draw up specifications for items such as the turbine generator and pollution control apparatus. The design contractor was then to integrate the power system, the research system, and the support system into the design for the total facility.

The second stage was to be the construction stage. During this stage the management was to evaluate the bids and to select an implementation contractor, who would handle all aspects of the construction. The contractor would be responsible for all subcontracting and for ensuring that construction meet design criteria.

The final stage was to be the operation and training phase. The plant would be operated by HELCO, which would contribute the time of a geothermal engineer to train a staff of technicians and power operators. After the facility was completed, the staff would operate the station and HELCO would purchase the electricity at the commercial price. In the proposal, it was estimated that the plant would have a yearly income of $260,000, which was enough to pay for all operational expenses plus leave a sizeable surplus.

Cost estimates of the proposal

Four cost options were included in the proposal. 1) The first option, costing $6,447,000, was for the basic facility as described, using the most modern equipment. 2) The second option, costing $5,189,000, was for the basic facility, but using a surplus Westinghouse turbine generator adapted to geothermal requirements. 3) The third option assumed the use of the surplus generator and also deleted the reinjection well from the basic facility. It was estimated to cost $4,655,000. 4) The final option assumed the use of the surplus generator, and deleted both the reinjection well and the research facilities. Summary of the costs for each option is in table 3.7.

Table 3.7. Hawaii Geothermal Research Test Facility Estimated Plant Equipment Costs (in Thousands of Dollars)

	Item	I	II	III	IV
1.	Turbogenerator	1,162.0	200.0	200.0	200.0
2a.	Substation	445.7	445.7	445.7	445.7
2b.	Instrumentation	126.2	126.2	126.2	126.2
2c.	Load bank	54.6	54.6	54.6	54.6
3.	Cooling water circulation system	54.3	54.3	54.3	54.3
4.	Steam separator	71.3	71.3	71.3	71.3
5.	Demister	16.0	16.0	16.0	16.0
6.	Hydrogen sulfide abatement	150.0	150.0	150.0	150.0
7.	Reinjection equipment	43.8	43.8	10.0	10.0
8.	Trailers (2)	61.9	61.9	61.9	61.9
9.	Condenser/eductors	61.5	61.5	61.5	61.5
10.	Cooling tower	87.5	87.5	87.5	87.5
11.	Site preparation	309.2	309.2	309.2	222.2
12.	Foundations	128.4	128.4	128.4	105.0
13.	Field piping	353.3	353.3	312.1	189.0
14.	Remote instrumentation				
15.	Injection well	428.7	428.7	107.6	107.6
16.	Research facility	866.7	866.7	866.7	0
17.	Miscellaneous	226.6	177.3	156.9	100.6
	Design costs	689	684	666	544
	Construction costs	1111	869	770	493
Total		6447	5189	4655	3100

Source: HGP-A Development Group, Proposal for a Geothermal Electric and Nonelectric Research Facility Utilizing the HGP-A Well on the Island of Hawaii, vol. 2 appendix A.

Project completion

While the consortium was negotiating with ERDA during 1977, HGP continued with its own project responsibilities, which increasingly centered upon the analysis of the data and the completion of final research reports. The environmental/socioeconomic program was the first to complete its research, formally concluding operations in January 1977 with the publication of a prototype environmental impact statement (EIS), entitled "An Assessment of Geothermal Development in Puna, Hawaii." Although the assessment was not represented as an EIS, it did contain much of the information that would be required in a formal statement. It contained, for example, the environmental baseline measures of the chemicals in the Puna area's air, water, and soil. It also described the plants and animals indigenous to the area, noting especially the rare and endangered species. Further included in the assessment was a discussion of the socioeconomic conditions in Puna, including the residents' employment pattern, the housing situation, the life-style, and the distribution of the population. Finally, the assessment compared potential benefits of geothermal development – such as implementation of an alternative form of energy, creation of more jobs in the area, and utilization of an indigenous source of energy – with potential costs, such as the transformation of agriculturally zoned land to industrial land, an increase in the noise level, and an increase in the amount of airborne pollutants. Overall, it was estimated that the benefits of geothermal development were substantially greater than the costs.

As previously described, the engineering team completed the well testing in May 1977. During the remainder of 1977, the engineering team members analyzed the test results and then, in 1978, published a summary reservoir engineering report.

Researchers in the geosciences program continued to analyze data from the earlier geophysical surveys and completed two new research tasks. In January 1977, the researchers conducted seismic refraction surveys, and in June 1977, they completed preliminary geochemical and hydrological analyses of the samples from the well flow tests. During the rest of the year, analysis of all geophysical and geochemical data continued, and attempts were made to integrate all data into a unified interpretation. Although the synthesis could not be accomplished during 1977, several research reports were sent to the Department of Energy in 1978 to summarize the geosciences work. These summaries included reports of all the geothermal explorations conducted from 1973 to 1977, including the electrical, magnetic, gravity, geochemical, seismic, and photogeologic surveys.

The initial plan had been to phase out HGP upon the submission of the research reports. However, the future of HGP depended upon the HGP-A Development Group proposal submitted to ERDA in April 1977. ERDA itself was being reorganized into the United States Department of Energy (DOE) during 1977, and after the transition was completed, a decision would be made. ERDA formally became the DOE in October

1977 and in November DOE officials notified the development group that DOE would fund the proposal. Specific details of the proposal would have to be worked out in negotiations with DOE.

Regardless of the future of HGP, the responsibility for the development of geothermal energy in Hawaii was now transferred to the HGP-A D/G, which was the entity that would accomplish HGP's ultimate goal of utilizing geothermal energy in Hawaii.

PHASE 4: EVALUATION AND REFINEMENT

No formal evaluation of the Hawaii Geothermal Project was conducted. This was because HGP was successfully integrated into the HGP-A Development Group and therefore became a part of the overall effort to develop geothermal power in Hawaii. From this perspective, the project is ongoing and cannot be evaluated independently of the larger development effort which has yet to be completed. In this section, then, a summary of the major results of the HGP between 1973 and 1978 will be presented.

Well test results

From the standpoint of accomplishing the implicit goal of discovering an exploitable geothermal resource, HGP proved to be a successful project. The HGP-A well was discovered to be one of the hottest geothermal wells in the world, with downhole temperature reaching 676°F (358°C). See figure 3.15 for a temperature profile of the well. There was a natural two-phase flow into the well bore with quality geothermal fluid and a substantial total flow rate. (See table 3.8 for a complete statistical profile.)

Table 3.8. HGP-A Discharge Results

	November	December	January	March
Well-head pressure (psig)	47	53	59	59
Well-head temperature (°C)	146	150	151	153
Mass flow rate (klb/hr)	88	103	114	120
Steam flow rate (klb/hr)	60	64	72	75
Steam quality* (percent)	68	62	63	62
Electric power potential (mWe)	3.4	3.8	4.3	4.5

*steam fraction.

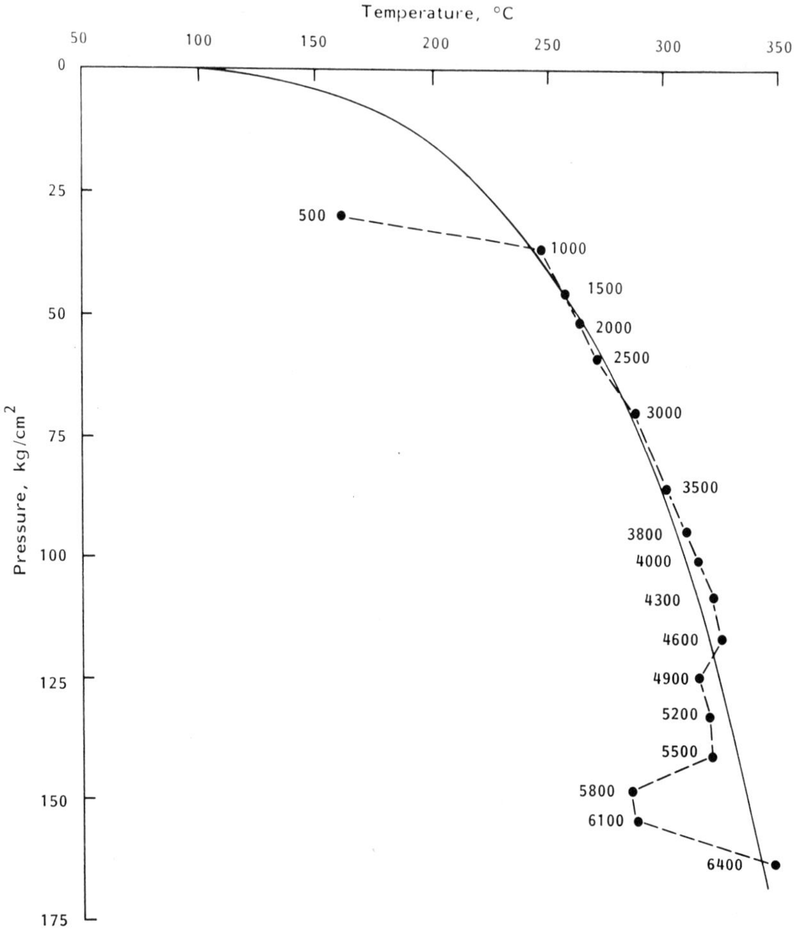

Fig. 3.15. Temperature profile of HGP-A well.

Based on the data collected from the well tests, it was estimated that HGP-A could generate 3 MW of electricity for a 30-year period. Table 3.9 shows the power projections for the well. It was also estimated that the entire reservoir feeding the well could be substantially larger. Estimates of the reservoir's generating capacity ranged up to 500 MW of electricity for the next 100 years. Compared to the island of Hawaii's total need of about 90 MW of electrical capacity and the state of Hawaii's present electrical capacity of 1400 MW, this was quite substantial. However, estimates based on a single well were not sufficient to accurately predict the capacity of the geothermal reservoir; it would be necessary to drill other wells for more information.

Table 3.9. Long-range Power Projections for HGP-A

Time in Years	Total Mass Flow Rate in klb/hr	Steam Flow in klb/hr	Well-head Pressure in psig	Enthalpy in Btu/lb	Power in MWe
1	81	59	153	900	3.2
15	78	58	142	904	3.0
30	77	57	140	906	3.0

Source: Hawaii Geothermal Project, HGP-A Reservoir Engineering, September, 1978, p. 42.

Geological results

Geologists and geochemists analyzed the cores and cuttings collected during drilling and found that the rock formation was tholeiitic basalt which could be divided into three zones of alteration (see table 3.10). Zone 1 where the alteration began, occurred between 2,220-4,265 feet (673-1,300 meters) and was characterized by montmorillonite, with minor calcite, quartz and chlorite. Zone 2 occurred between 4,455-6,250 feet (1,350-1,894 meters), with the principal alteration mineral being chlorite and accessories being quartz, actinolite, and montmorillonite. The boundary temperature between zones 1 and 2 was about 617°F (325°C). The third zone became dominant from about 6,234 feet (1,900 meters) to the bottom of the well. Actinolite predominated in this zone with chlorite, quartz, pyrite, and hemitite secondary. The boundary temperature between zones 2 and 3 was 644°F (340°C).

From the top of the well to a depth of about 3,500 feet (1,067 meters) the lava was highly permeable, with excellent permeability between 2,500-3,000 feet (762-914 meters). Then, from about 3,500-6,200 feet (1,067-1,890 meters), the permeability became poor, although layers of medium permeability existed throughout the dike making possible geothermal production. At the bottom of the well, from 6,200-6,600 feet (1,890-2,012 meters), the permeability was excellent. Figure 3.16 illustrates the zones of permeability as they relate to the HGP well.

Based on this information, the HGP researchers derived several speculative models of the well's underground system. One of the most probable models depicted two production zones, one at about 4,400 feet (1,341 meters) and the other at about 6,400 feet (1,951 meters). Both zones were supplied by aquifers that were recharged by rainfall percolating into the ground. However, only high levels of rainfall could penetrate to these zones because of the alternating layers of poor permeability. The heat at these depths was sufficient to boil the water and produce steam.

Table 3.10. Geologic Analysis and Interpretation of Possible Production Zones

Depth from Well-head (feet)	Microscopic Analysis	Megascopic Analysis	Boundary Temperature Zones of Alteration
0			
500	Unfilled	Generally high permeability	Little or no alteration
1,000	Unfilled		
1,500	Unfilled		
2,000			
2,500	Partially Filled		
		Higher Permeability	Zone 1 Major mineral: montmorillonite Minor minerals: chlorite, quartz, calcite
3,000	Partially Filled		
3,500	Filled	Generally low permeability, but possibility for layers of medium permeability	
4,000			320°C
4,500	Filled		
5,000			
		Varying but generally low permeability	Zone 2 Major mineral: chlorite Minor minerals: quartz, actinolite, montmorillonite.
5,500	Filled		
6,000	Filled		
			340°C
6,500	Partially filled	High permeability	Zone 3 Major mineral: actinolite Minor minerals: chlorite, quartz, pyrite, hemitite

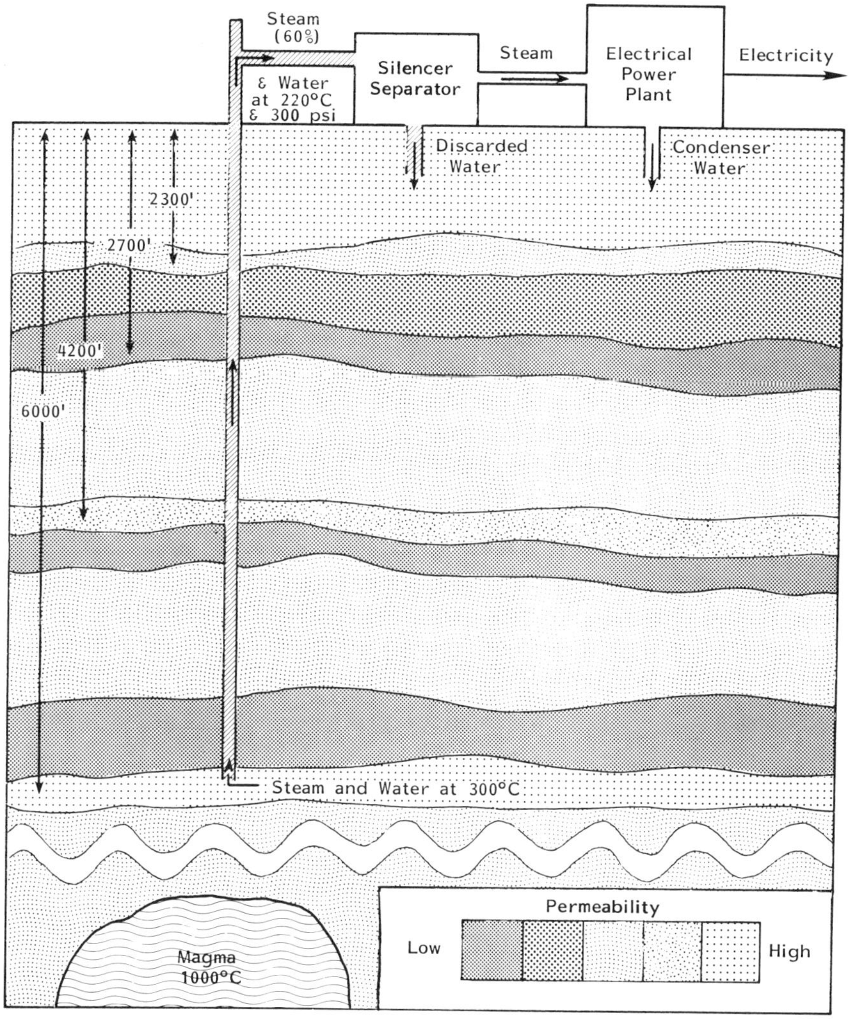

Fig. 3.16. Layers of permeability as related to HGP-A.

Well emissions

During the well testing, the HGP researchers collected atmospheric samples to determine the airborne chemicals emitted from the geothermal discharge, and they collected downhole samples to analyze the chemicals in the geothermal fluid. With respect to airborne emissions, there was concern about sulfur dioxide, hydrogen sulfide, and mercury. Both sulfur dioxide and hydrogen sulfide were bothersome, but the levels were never dangerous. In fact, their levels in the atmosphere did not change during the well tests, thus indicating that the well was not adding significantly to the natural volcanic emissions of sulfide into the atmosphere (see table 3.11). However, the odor of hydrogen sulfide is particularly offensive to humans; it smells like rotten eggs. Moreover, it can be detected in quantities as small as 3 parts per billion. Thus, although the flashing of the well did not significantly increase the atmospheric levels of hydrogen sulfide, it did disturb nearby residents. This was especially true when there was no wind or when the prevailing wind blew toward nearby homes. The last bit of odor would have to be eliminated before further developments could take place.

Table 3.11. Aerometric Data for HGP

Date	Well Status	Sulfur Dioxide*	Hydrogen Sulfide*	Mercury**
May 1975	Predrilling	0.5	0.5	1.1
May 1975	Predrilling	0.5	0.2	1.2
July 1976	Flashing	0.5	0.2	9.9
Nov. 1976	Well shutdown	–	–	0.1
July-Aug. 1977	Well shutdown	0.3	0.2	0.8

*Parts per million.
**Microgram per cubic meter.
Source: Adopted from, Revised Environmental Impact Statement for the Hawaii Geothermal Research Station Utilizing the HGP-A Well at Puna, Island of Hawaii.

Levels of mercury were another matter. Mercury is potentially dangerous at any level and the provisional federal standard is not more than 0.1 microgram per cubic meter for continuous exposure. Mercury levels near the drill site were high, and during the July 1976 flashing, mercury levels rose to 9.9 micrograms per cubic meter. Later analysis, however, revealed that the area naturally contained high levels of atmospheric mercury. Moreover, the high levels recorded in July were caused not by the well testing, but by volcanic vents. Table 3.11

illustrates that even with the well shut, mercury in the air sometimes exceeded 0.1 micrograms per cubic meter. After monitoring the air for two years, researchers could find no evidence of a buildup of mercury.

The chemical content of the geothermal fluid was a cause of concern because any geothermal plant would have to dispose of or reinject the used geothermal fluid back into the earth. This could contaminate the ground water. However, the chemical content of the HGP geothermal fluid did not differ significantly from that of the brackish water wells in the area (see table 3.12). This suggested that the area's ground water had been naturally contaminated due to the upward movement of heated salt water and that no Ghyben-Herzberg lens – a pool of fresh water floating on salt water – existed in the area. Only one potential hazard existed, silica. Because of the high downhole temperature, the level of silica in the HGP fluid was 440 milligrams per liter, several times higher than normal. It would have to be filtered out before the well's geothermal fluid could be reinjected into the ground.

Table 3.12. Comparison of Chemical Content of HGP-A with Nearby Wells and Springs (Milligram/Liter)

Site	Chlorine	Calcium	Potassium	Magnesium	Sodium	Silicon Dioxide
HGP fluid (downhole)	925.0	84.2	135.0	2.1	830.0	440.0
Isaac Hole Spring	3,534.0	32.4	86.0	200.0	2,020.0	81.5
Airstrip Well	303.5	23.0	13.6	28.0	238.0	71.3
Allison Well	281.0	13.4	10.8	15.0	216.0	24.1
Malama Ri Well	3,811.0	66.8	109.0	210.0	2,105.0	81.5
Rain water	7.2	0.25	0.25	0.75	4.5	0.0

Source: Adopted from Hawaii Geothermal Project, Environmental Baseline Study for Geothermal Development in Puna, Hawaii, September 1976; and Hydrology and Geochemistry of Hawaiian Geothermal System: HGP-A.

Noise

Despite constant improvements in the muffling of the well, noise continued to annoy residents throughout the testing period. Although the noise did not exceed Environmental Protection Agency standards when measured at nearby houses, it was high pitched and thus irritating. Moreover, many residents were accustomed to the quiet of the rural area, while others had moved to Puna specifically to get away from the loud noise. Table 3.13 provides data on the noise. Thus any subsequent developments would have to eliminate the noise problem.

Table 3.13. Level of Noise near Well Site
(in Decibels)

Location	7/19/76 (without Silencers)	11/3/76	1/27/78	2/10/77 (3-Inch Orifice)	5/7/77 (3-Inch Orifice)
Corners of well					
40 feet away	113	100	96	98	–
40 feet away	113	98	93	95	91
70 feet away	94	98	89	91	–
70 feet away	94	98	91	91	89
At roadside, 100 feet away	122	87	80	80	75
Estimated					
1 mile away	–	–	–	–	70
2 miles away	–	–	–	–	49

However, HGP was only an experimental program; the demonstration plant had not been built. When electricity was actually generated, the sound would be muffled by the generator, into which the steam would be fed, and by the building, which would house the generator and the operating facilities. It was anticipated that this would reduce the noise to an acceptable level.

Economic aspects of the project

The Hawaii Geothermal Project was from the beginning a cooperative effort among the federal, state, and county governments, the utility companies, and the University of Hawaii. Each contributed by giving the project expert advice and by providing services when appropriate. In addition, each group supported the project financially. During the years 1973 to 1978, HGP's total allocation amounted to $3,387,000. Table 3.14 illustrates the total funds granted to the project during this period.

Although this total amount was large, it must be put into perspective. First, the total was spent over a period of five years, and it

Table 3.14. Financial Summary for HGP, 1973-1978
(in Thousands of Dollars)

Funding Entity	1973	1974	1975	1976	1977	1978	Total
National Science Foundation	269	200					469
Energy Research and Development Administration/Department of Energy (1977)			119	1472	270	147	2,008
State of Hawaii	100		500		66		666
County of Hawaii	100						100
Water Resources International				60			60
Hawaiian Electric			45				45
Other			39				39
Totals	469	200	703	1,532	336	147	3,387

funded numerous activities ranging from geophysical surveys to socio-economic assessments. Second, the project was intended to provide the basic research and development that would lead to the exploitation of geothermal resources in Hawaii. It was intended neither to be an exploration for geothermal resources nor to be an eventual profit-making venture. Finally, the project did discover a productive geothermal well and a potentially large geothermal reservoir. It is this well and potential reservoir that the 3.5-MW demonstration plant will utilize.

EPILOGUE

In June 1978, negotiations between the Department of Energy and the HGP-A/DG were completed and a four-year well-head generator contract was signed providing $6,268,256 of local and federal support. An additional agreement was reached with the utility company, which would purchase an estimated $482,758 of electricity generated by the well during the initial two-year period of operation. Design of the generator system is nearing completion, and the facility will begin supplying electricity to the residents of Hawaii in early 1981.

NOTES

(1) City and County of Honolulu, Report of the Mayor's Energy Resources Committee, 1973, p. 1.

(2) Robert M. Kamins, An Assessment of Geothermal Development in Puna, Hawaii, Hawaii Geothermal Project, January 1977, pp. 4-5.

(3) Department of Planning and Economic Development, State of Hawaii, "Energy Use in Hawaii," in The Role of Alternate Energy Resources in Promoting Island Self-Sufficiency, Hearings before the Subcommittee on Energy Research and Development, (Washington, D.C.: U.S. Government Printing Office, 1979), pp. 110-111.

(4) Kamins, An Assessment of Geothermal Development, p. 5.

(5) Center for Engineering Research, University of Hawaii, A Proposal to the Hawaii State Legislature in Response to House Resolution 351 Requesting a Study of New Energy Sources for the State of Hawaii, 1971, p. 7.

(6) Honolulu Star-Bulletin, March 5, 1973, p. A-20.

(7) Hawaii Geothermal Project, Summary Report for Phase I, May 1975, p. 27.

(8) Honolulu Advertiser, February 9, 1974.

(9) A.S. Furumoto, G.A. Macdonald, M. Druecker, and P.F. Fan, Preliminary Studies for Geothermal Exploration in Hawaii, 1973-1975, December 1977, Hawaii Institute of Geophysics, University of Hawaii, p. 14.

(10) Hawaii Geothermal Project, Phase II Revision to Proposal AER7500285-000, December 1974, pp. 2-51.

(11) Ibid., pp. 5-6.

(12) Hawaii Geothermal Project, Summary Report for Phase I, p. 27.

(13) Kingston Reynolds Thom and Allardice Limited, Hawaii Geothermal Project Testing Program, August 1975, pp. 1-2.

(14) Kingston Reynolds Thom and Allardice Limited, Hawaii Geothermal Project Drilling Plan, p. 2.

(15) Entry in the drilling log of the Hawaii Geothermal Project, n.d.

(16) Personal communication from John Shupe.

(17) Personal communication from John Shupe.

(18) It was decided that if the well was successful it would be named in honor of Agatin Abbott, the initial drilling program coordinator, who died on July 31, 1975. Hence the "A" in HGP-A stands for Abbott.

(19) Honolulu Advertiser, April 22, 1977, B1.

(20) Ibid., B1.

(21) Hawaii Geothermal Project, Environmental Baseline Study for Geothermal Development in Puna, Hawaii, September 1976, pp. 5-7.

4 TIWI Geothermal Project: The Philippines

Rosemary Aquino and Salvador Aquino
with the technical assistance of
Arturo Alcaraz

PROJECT BACKGROUND

The Philippines is an archipelago of 7,100 islands located in Southeast Asia. Its land area occupies approximately 115,000 square miles (297,850 kilometers), its island waters 600,000 square miles (15,540,000 square kilometers). Eleven islands account for 92 percent of the total land area. Ninety-six percent of the population lives on these main islands. The population in 1960 was 27 million, growing at an average rate of 3 percent.

There are three geographical regions in the country: Luzon, Visayas, and Mindanao. Luzon has an area of 54,000 square miles (139,860 square kilometers) and is the largest region in land area and most important in terms of economic activity. The capital and principal commercial city of the country, metropolitan Manila, is located in Luzon. The Visayas region occupies 25,000 square miles (64,700 square kilometers) and is located between Luzon and Mindanao. The region is dotted with numerous islands and waterways. The Mindanao region, containing 37,000 square miles (95,830 square kilometers), is southernmost in location. Dubbed the agricultural land of promise, the region has numerous agricultural plantations as well as industrial sites.

The economy of the country is based on agriculture. In the early 1960s, this sector accounted for over one-third of the net domestic product (NDP). Over half the working population is employed in this sector. Industry is a rapidly growing sector, with a share of NDP slightly below 20 percent. The government supports and encourages the industrialization of the country by providing basic infrastructure and incentives.

The country, as an old legend goes, has been blessed by the gods. There are numerous resources available – from mountains and minerals to lush tropical forests, waterways, and marine life. Many of the raw material resources are virtually untapped. Because the Philippines is a developing country, it is progressively harnessing its potential in manpower and natural resources to attain national objectives.

In terms of energy resources, the Philippines is also fortunate. In the early 1900s, coal extracted locally was used as a source of energy to fire kilns for industries such as cement and transport. Firewood from trees in the forests was used in households for cooking purposes. A government agency, the National Power Corporation, was established to develop the hydro resources as sources of power.

By the mid-1900s, however, due to the availability of cheap oil from foreign sources, the country gradually shifted to the use of oil as a source of energy. By the early 1970s, more than 90 percent of the country's energy requirements were met with the use of imported petroleum fuel.

Before the energy crisis of 1973, government officials in the Philippines focused attention on developing local energy resources. The government program for the development of rural areas relied, to a large extent, on the availability of indigenous, localized (especially for island provinces), and possibly cheaper sources of power for the industrialization and infrastructure building needs. One of the energy resources chosen for commercial exploitation was geothermal energy, which was considered potentially available since the Philippines is located in the so-called circum-Pacific "rim of fire." The country is in the volcanic zone and has a total of 50 volcanoes punctuating the length of the archipelago; 12 of the volcanoes are active.

Numerous surface manifestations of geothermal energy, such as hot springs, have been evident for many decades. These hot springs have popular appeal to local residents, who regard them as having curative properties. Resorts are often built around the springs, and the name bano, meaning bath, is frequently used. The waters are said to heal muscular ailments, rheumatism, respiratory and skin diseases. Many tourists find the waters relaxing to soak in.

PHASE 1: PLANNING, APPRAISAL, AND DESIGN

Preliminary Research Studies

Geothermal energy has always been linked with volcanic activity, since geothermal resources are reservoirs of boiling water and steam found in high heat flow regions. Initial data needed for geothermal exploration include: geological investigations of volcanic areas; data on surface manifestations such as bubbling mud or water pools, steam vents, hot springs, and geysers; and chemical analyses of volcanic rocks and hot spring waters. These data are often gathered by government agencies doing work related to volcanic studies.

In the Philippines, because of the very real danger from volcano eruptions, several government agencies monitored volcanic activity. The Weather Bureau made seismic studies. The Bureau of Mines, previously the Division of Mines, studied the geologic and economic aspects of vulcanism. The Bureau of Science, now the National Institute of Science Technology, performed chemical analyses on rock, gas, and water specimens. The Department of Geology and Geography of the University of the Philippines assisted in the geological investigations of volcanic areas in the country.

The first recorded study of thermal springs in the Philippines was a survey undertaken in 1926. In 1928, Dr. Jose Feliciano, head of the Department of Geology and Geography of the University of the Philippines, compiled a list of 54 thermal springs located throughout the country. The list indicated the measured temperature and partial chemical analysis of the springs.

Not much more research on vulcanism and surface manifestations was published until 1951. In 1951, the consolidation of the volcanologic studies was started with the formation of a committee on volcanology under the Department of Agriculture and Natural Resources. The members of this committee were representatives of the government agencies involved in volcanologic studies.

In 1952, after the destructive eruption of the Hibok-Hibok Volcano in Camiguin Island, Mindanao, which claimed 500 human lives, a congressman from the area, Emmanuel Pelaez of Misamis Oriental, proposed a law creating the commission on volcanology. The president of the Philippines subsequently signed into law Republic Act No. 766, which created the Commission on Volcanology to take over the functions of the committee on volcanology. These functions were: to observe all active volcanoes in the country, predict eruptions, and plan evacuation and relief programs. The Commission on Volcanology (COMVOL) was placed under the National Research Council of the Philippines (NRCP) and was designated the lead agency for research on volcanoes and volcanology.

The COMVOL program, set up with the assistance of Dr. Gordon A. Macdonald, director of the Hawaiian Volcano Observatory, included geological and topographic investigation, seismological and other geophysical investigation, strictly volcanological investigation, and warning and relief organization. Research studies made during the next decade were the completion of a bibliography of literature on Philippine volcanoes, papers on volcanology, and detailed geological surveys of critical areas in the vicinity of active volcanoes and solfataric areas.

Interest in research on geothermal energy started when the COMVOL office received a copy of the proceedings of the Rome United Nations Conference on New Sources of Energy in 1962. One of the findings at this conference was that geothermal energy was the cheapest means for generating commercial electrical power. Because of the interest of the COMVOL geologists in this finding, the COMVOL chief volcanologist directed field teams to start taking temperature measurements and some chemical analyses of volcanic rocks and hot spring waters in areas which had surface manifestations such as in Los Banos, Laguna and Tiwi, Albay.

In the meantime, in the decade of the fifties, the Philippine government completed the reconstruction and rehabilitation of the country which had been ravaged during World War II. The focus by the 1960s was on the development of industry in the urban centers and in the countryside.

Electric power was considered a critical element in the industrialization effort. Hence, the government commissioned a study to examine the power sources and needs for the main Philippine region, the Luzon area. In the early 1960s, the main sources of power in Luzon were three major hydroelectric plants producing 207 MW of electricity, as well as imported petroleum to fire thermal plants. There were limitations to the use of each type of energy. Use of imported oil depleted the country's foreign exchange reserves, while hydroelectric projects, especially in Luzon where a distinct wet and dry season exists, were affected by low water levels during the dry season.

In 1963, the United Nations Special Fund provided the assistance for the study on power which was to be executed by the International Atomic Energy Agency (IAEA). The United Nations study, entitled "Pre-Investment Study on Power, Including Nuclear Power in Luzon," was to evaluate the resource potential in the country of water or hydro, coal, geothermal, oil, and gas. The studies were to be made separately by experts in the respective fields. If these resources were found to be insufficient to meet the demand, nuclear energy plants would be recommended to meet the deficiency.

The International Atomic Energy Agency contracted G.W. Grindley, a geothermal energy consultant from New Zealand, to prepare the geothermal study. For this purpose, he visited the Philippines during December 1963 and January 1964, and conducted field research with the help of Bureau of Mines and COMVOL personnel. He even experienced the heat of the Tiwi geothermal area first hand, having scalded a leg in the process of field study. His final report, written in February 1964, was entitled, "Geothermal Resources of the Island of Luzon, Philippines."

The geothermal report discussed the nature of geothermal energy, reviewed existing information on geothermal areas in the country, indicated the areas in Luzon with the most promising geothermal resources, and provided a technical guideline based on New Zealand experience for local geologists to follow in carrying out geothermal surveys.

In his report, Grindley confirmed that the Philippines was well endowed with geothermal resources, since the country was located in a high heat flow region with active, dormant, and extinct volcanoes, and had a number of surface manifestations in the form of hot springs, fumaroles, and solfataric areas. Figure 4.1 is a map showing the location of volcanoes and hydrothermal fields, and indicates the close relationship among the majority of them.

The paper noted that some studies had been started on the chemical analyses of the hot springs. No exploratory drilling, however, had been done so far to determine the potential and characteristics of the geothermal fields. Consequently, the report focused more on findings based on the study of the hot spring waters and the geology of the areas.

A total of 66 hydrothermal fields was listed, and chemical analyses were made on the majority of the springs. The springs were classified,

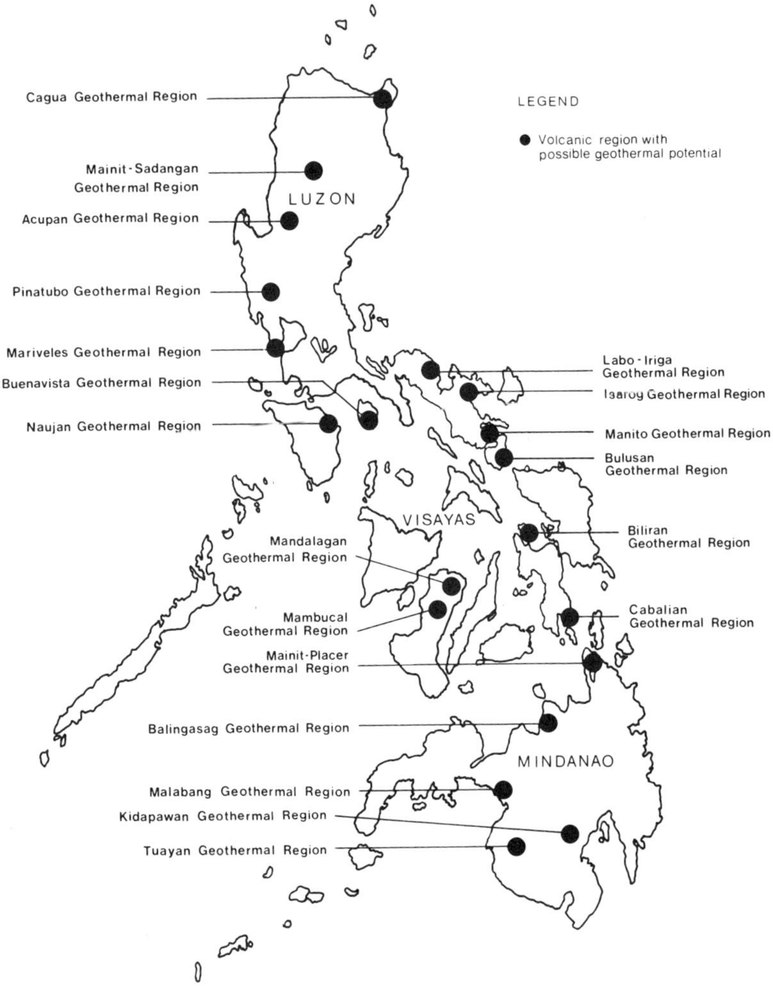

Fig. 4.1. Location of geothermal regions in the Philippines.

on the basis of their chemical content, into: those with sodium chloride waters (15); those with acid sulfate-chloride waters (5); those with calcium-sodium chloride waters (4); and those with dilute waters (24).

The hot springs were further divided into four broad classes based on geological association, and the report noted examples of each type of spring based on preliminary studies of the waters and geology.

Class 1. direct volcanic exhalations

These are the hot springs found in or near the craters of active volcanoes. Such springs usually have a major volcanic component with varying amounts of contamination by meteoric waters. Because of this circumstance, the springs are usually of little economic interest for geothermal energy development. An example is the Taal Volcano in Luzon.

Class 2. shallow volcanic sources

These are the hot springs found around the margins of many active and recently extinct volcanoes. They derive their heat supply from the downward penetration and subsequent uprising of meteoric water to the hot rock at the roots of the volcano. Since the volume of hot rock is not usually very large and cools rapidly close to the surface, such springs may not rise notably in temperature with depth. The Los Banos spring is an example of this type.

Class 3. deep-seated plutonic sources

These are extensive thermal areas with a large heat flow which are fed from deeper plutonic sources, probably cooling magma chambers or regions of deep-seated granitization. Large magma chambers within the crust are apparently formed by large-scale crustal melting, and are a feature of many acidic volcano provinces. The capacity of the heat source in such areas may be extremely large, and where permeable near-surface formations allow deep penetration of meteoric waters, large convection systems may be set up. Temperatures in such systems may be high (200°-300°C, or 392°-572°F), the volumes of hot stored water may be large, and the surface heat flow considerable, such as in Wairakei, New Zealand. An example in the Philippines is the Tiwi Field in Albay Province, which is fed from a major fault, possibly tapping a plutonic source.

Class 4. deep faulting in region of high geothermal gradient

These are hot springs not directly related to volcanic or plutonic activity. Volcanic provinces and mountainous regions of recent rapid tectonic uplift commonly have a steep geothermal gradient, and deep faulting in such areas may be the cause of the formation of hot springs. An example of this type of hot spring is that found in the Mountain Province of Northern Luzon.

Based on his observations of the hot springs and the geology of the volcanic and solfataric areas, Grindley in his report identified Tiwi, Albay Province in the Bicol Peninsula as the most promising geothermal area. Among the reasons he cited were:

1. The area is associated with a major fault, possibly tapping a plutonic source. It is apparently unrelated to nearby shallow volcanic sources.
2. The country rocks are andesitic lavas and tuffs, of late Tertiary and Quarternary age, that are likely to contain permeable horizons.
3. The heat flow at the surface is of the order of 10,000 kilocalories per second (kcal/sec). This is approximately one-tenth the heat flow for the Wairakei Field in New Zealand.
4. Practically all the hot water discharged at Tiwi is at or near boiling temperatures. Although the rise of temperature with depth is not known, boiling point conditions are probable for at least 100 feet (30.4 meters).
5. The water being discharged by the Naglagbong springs at Tiwi contains a high concentration of sodium chloride, about four times that of Wairakei. Since the springs are close to the sea and only 40 feet (12.2 meters) above sea level, an admixture of meteric water and sea water could account for the high salt content. Since the interface between cold fresh water and cold sea water should be about 1,600 feet (488 meters), or (40 x 40, or 12.2 x 12.2), the convection system feeding the Naglagbong springs must extend to at least this depth.
6. Although beyond the present range of the main power network serving the Luzon region, the Tiwi Hydrothermal Field is situated in a densely populated area where alternative sources of power are lacking. This meant that demand for power existed near the source of geothermal heat.

Grindley recommended a study of possible uses for this geothermal energy. Electric power generation was definitely one use if the minimum temperature were no lower than 200°C (392°F), a temperature considered desirable by the New Zealand engineers for the exploitation of the geothermal energy to be commercially feasible.

Another use he suggested was the drying of copra, abaca (hemp), fruit, rice, vegetables, milk, and other commodities, especially in high rainfall areas where sun drying was commonly impossible. Geothermal heat, he noted, was also used in New Zealand for timber drying and salt making.

Another recommendation was that more surveys should be carried out by field personnel of the COMVOL and the Bureau of Mines over a few years' period in all the known hydrothermal fields. The purpose of the surveys would be to come up with more data on the temperatures and heat flow, geological structure, and chemical analyses of the thermal waters. These data would establish the scientific basis for the utilization of geothermal energy.

After the geothermal report was concluded, the COMVOL and the Bureau of Mines added the survey activities to the regular mapping and volcanological programs of their field teams. The COMVOL intensified its geothermal studies since these years were quiet years with no major volcanic eruptions, thus allowing the geologists to devote time to the project.

Arturo Alcaraz, then chief volcanologist of the COMVOL, set up the program of activities of the COMVOL in 1964 to include:

1. Preparation of a comprehensive list of existing hot springs, to be accomplished with the help and cooperation of the local and provincial authorities.
2. Preliminary geothermal studies of eight areas identified by surface manifestations.
3. Detailed geologic survey of each thermal field, to establish as far as possible the lithology and structure, and to locate faults and other structures with which thermal fields may be associated.
4. Measurement of the surface thermal gradient, to establish the limit of warm ground of each thermal area and whether they are connected by a regional belt with a thermal gradient higher than normal.
5. Magnetic survey of each thermal field, to determine magnetic properties of rocks due to hydrothermal alterations. This information would be helpful in locating areas of good potential.
6. Collection and chemical analyses of gas and water samples, to determine the chemical composition, which can give some ideas of the origin of thermal waters.
7. Drilling operations to determine the rock structure and temperature below the surface which can be correlated with other data gathered.

Whenever possible, through grants or scholarships, the COMVOL scientists were encouraged to learn more about geothermal energy.

Since the COMVOL was a research-oriented government agency that was allocated a yearly budget by the Congress of the Philippines, Alcaraz requested the Budget Commission and the Congress to allocate to the COMVOL an additional ₱100,000 to fund these studies. This budget was subsequently approved.

Alcaraz next approached the executive director of the National Science and Development Board (NSDB), a government agency that funded local scientific studies, for the funding of a geothermal research project at Tiwi. Two research projects were eventually approved. The first was "Project 2.91: A Study of the Tiwi Geothermal Area for Power Utilization"; and the second was "Project 2.186: A Study for Commercial Development of the Tiwi Geothermal Field for Power and Industrial Purposes." The amount of ₱311,000 was allocated for the first project, in 1964; ₱813,564 for the second project in 1967.

The objective of the first project, Project 2.91, was to study the extent of the Tiwi thermal springs and their heat output and to make pilot studies on the economic utilization of the geothermal heat. The project was carried out with the help of the Bureau of Mines, which would assist in exploratory drilling. The project consisted of three phases:

Phase 1: A detailed geologic and geophysical study as well as exploratory drilling of the area.
Phase 2: Drilling and logging of drill holes.
Phase 3: Construction of a pilot power plant.

The research project was completed successfully in 1967. During the course of the research, some significant findings were:

1. The Tiwi thermal field is located in a valley that is covered mostly by Quaternary alluvial materials associated with some undifferentiated talus deposits. The alluvial fan is perhaps dominantly underlain by the pervious Malinao (volcano) lavas dipping toward the valley floor. Somehow, a major part of the thermal water from the hot springs may be attributed to the presence of this artesian structure.

2. The Tiwi thermal field is divided into four thermal areas: Naglagbong, Bano, Nahologan, and Cararayan. Thermal activities are mostly confined within the barrios of Naglagbong, Bano, and Nahologan, where the Tiwi arcuate fault conjugated by younger minor faults passes through. The Naglagbong has the widest area of thermal springs, while Bano has the smallest. The Naglagbong, too, exhibits the highest average surface temperature while Cararayan exhibits the lowest.

	Naglagbong	Bano	Nahologan	Cararayan
I. Area of thermal field (square miles)	28.1×10^4	10.2×10^4	18.3×10^4	12.9×10^4
II. Average surface temperature range in C^o (F^o)	60-103 (140-217)	40-100 (104-212)	35-95 (95-203)	40-65 (104-149)

3. Field mapping and aerial photographs done with the help of the Philippine Air Force led to the discovery of three fault systems. Most of the faults in the thermal area were inferred and plotted from the alignment of hot springs. The fault systems detected were as follows:

a. The Plug-Dome Fault System: The faults that belong to this system are located on the eastern boundary of the thermal field. These are the older faults and in reality are in no way related to the thermal activities west of them. The fault system is more associated with the Bolo and Putsan plug-domes, for the faults that belong to this system were responsible for the formation (Lagonoy fault), and the shaping up of the present topography of the plug-domes. The most important of these faults is the Logonoy fault, the oldest structural feature of the whole area covered by the geologic map. The structure, a high-angle fault striking 325°, dips steeply to the northeast or toward the sea.

That the fault remained active after the extrusion of the hyalodacitic lava that built up the Putsan plug-dome is deduced from the nearly parallel orientation of the lavas close to the fault. It is supposed that the rotation of the mica flakes into an approximate parallelism was caused by a shearing movement along the Lagonoy fault.

The Lagonoy fault is cut by the almost east-west trending Cararayan fault at the contact between the two plug-domes. The dip and the displacement of this fault is not quite clear, although there is reason to believe that the northern block (the Putsan plug-dome) moved up with respect to the southern block (the Bolo plug-dome). In the north, the Lagonoy fault is displaced to the right by the Cararayan fault.

To the north of the Putsan plug-dome is the Putsan fault; to the south of the Bolo plug-dome is the scarp-forming Bolo fault, which has a strike of approximately 45° and a dip heading to the southeast. A small displacement on the Bolo fault was possibly caused by further renewed activity of the Lagonoy fault.

b. Radial Fault System: The radial fault system is typified by the Joroan, Matalibong, and Pili faults. Aerially, these faults originate from the vicinity of the volcano's crater and gradually diverge into nearly parallel lines to the northeast. These faults are almost vertical without any appreciable displacement. As in the plug-dome fault system, there is no field evidence that these faults are related to the thermal activities.

c. Arcuate Fault System: To this system belong the faults that originated as a result of the collapse of the northeastern face of Mt. Malinao (geologic setting). These are the Kagumihan fault and the Tiwi arcuate fault zone with its complement of small minor arcuate and cross faults. The Tiwi arcuate fault zone, the most striking and the longest fault in the area, is probably the deepest. It starts from the Bolo fault to the east, moving to the northwest to swing in a southwesterly direction toward Mt. Malinao.

It is strongly believed that the thermal activities have some definite relation or association with these faults, particularly the Tiwi arcuate fault zone. They are the most important structural features of the area, as far as thermal activity is concerned, for they serve as feeding channels for hot solutions to escape to the surface. Most of the hot springs within the Bano, Naglagbong, Nahologan, and Cararayan thermal areas apparently are related to the Tiwi arcuate fault zone. The fissures become a locus for thermal activity. Magmatic water and solutions from the cooling Malinao magma mingled with heated ground water gradually find their way to the surface to form thermal springs.

4. The source of heat in the Tiwi thermal field was not conclusively indicated in the geologic report. The possible explanations for how heat is transferred to the surface can be given as: convection, conduction, or condensation of the magmatic source.

Through the surface data gathered from the field, it was considered that the hot springs in the four thermal areas are fed chiefly by ground water, with the admixture of a small portion of magmatic water. The magmatic water that rises into the surface as steam associated with some volcanic gases, through crevices or cracks and deep fissures, is condensed by ground water and in turn mingles with it. If there is any difference in the amount of magmatic water in the different thermal areas, it may be the result of the variation in the supply of ground water in the different thermal areas as caused by the seasonal variations and stratigraphic structures, but not the constancy of magmatic emanations.

In view of these factors, the water coming out from the different thermal areas originates from a depth up to which ground water can penetrate, and the mineral contents besides the volatile constituents are primarily derived from the rocks above the level of the magmatic source.

5. Based on findings of the preliminary surveys, siting of exploratory drill holes for study of underground structure and temperature gradient were made. During the second year of the project, six shallow exploratory holes were drilled by the Bureau of Mines, with the use of a Longyear diamond drill. As the initial holes showed encouraging results, the holes were drilled to double the original 330-foot (91-meter) depth. Two drill holes showed geysering activity, while the last exploratory drill hole manifested continuous steaming with a well-head pressure and temperature of 115 psig and 99°C (210°F), respectively. The depth of this steam producing hole was 552 feet (167 meters) (see table 4.1).

6. A pilot plant was set up to demonstrate the possibility of utilizing natural steam to generate electric power. A steam calorimeter was used to study the quality and potential of the steam from the producing exploratory drill hole. Improvised separator, water tanks, and pipeline systems were installed. An Elliot turbogenerator, borrowed to test run the geothermal steam, ran successfully, producing a 3-kW power.

The second research project was started once the first project was successfully completed. The objective of this project, Project 2.186, was to obtain more information on the extent and potential of the Tiwi geothermal field for use by the government in the full exploitation of the field. Since the earlier project had accomplished the shallow drilling of exploratory drill holes and the results were favorable, a seventh drill hole was planned for a depth of 1,000 feet (305 meters).

Since neither COMVOL nor the Bureau of Mines had the equipment or experience for deeper drilling, the COMVOL contracted an American drilling contractor, Bezette Diamond Core Drilling Inc., to drill the seventh hole. Bezette used a Failing rotary drill rig to do the job.

COMVOL soon discovered, however, that the contractor could not cope with the job. The hole took four months to drill, including several weeks' delay due to various technical difficulties suffered by the contractor. Several blowouts occurred during the process of drilling. The 4-inch (10-centimeter) diameter drill hole was supposed to have been drilled 1,000 feet (305 meters), but the contractor gave up at 641 feet (195 meters) when the drillers encountered steam and could not continue due to a blowout with strong discharges of hot water and steam. This first production hole discharged 95 percent pure steam at

Table 4.1. Data on Drill Holes, Tiwi, Albay

Hole Number	Location	Depth – feet (meters)	Maximum Temperature at depth Indicated in feet (at Depth in Meters)	Casing–inches (cm)	Discharge
DDH-1	Visiting Naga	304 (93 m)	36°/304 (97°F/93m)	1 (2.54 cm)	Warm (30°C) 86°F water at 4-8 liters/min (1-2 gal/min)
DDH-2	Cale	612 (186 m)	132°/484 (270°/147m)	1 (2.54 cm)	Geysering of gas and hot water with a well-head pressure of 212 psi and 60°C (14°F), respectively
DDH-3	Nahologan	358 (109 m)	136°/315 (277°/96m)	1 (2.54 cm)	Geysering with a well-head pressure of 50 psi
DDH-4	Cararayan	494 (150 m)	42°/445 (108°/136)	uncased	Cold water without artesian pressure
DDH-5	Calzada	327 (100 m)	46°/327 (115°/100)	2 (5 cm)	Cold potable water discharging at 57 liters/min (15 gal/min)
DDH-6	Cale-Cararayan	552 (168 m)	over 129°/342 (264°/104)	2 (5 cm)	Steaming of high quality with a well-head pressure and temperature of 115 psi 99°C (210°F), respectively
DDH-7	Cale- Cararayan	641 (195 m)	unmeasured	6 (15.2 cm)	Almost pure steam with a well-head temperature of 154°C (309°F)

75 psig blowing from the 4-inch (10-centimeter) diameter outlet. The dry steam was discharged at the rate of 9.9 tons per hour, at a well-head pressure of 5 atmospheres and a temperature of 154°C (309°F) and the enthalpy at 1,160 Btu/lb.

COMVOL next engaged the services of another American firm, this time one which had actual experience in geothermal energy development. The firm, Rogers International of San Francisco, had participated in the geothermal power development of The Geysers, California. Rogers International and its Philippine subsidiary, Rinfil Engineering Corporation, were to assist in conducting instrumented steam-flow tests on the first production hole.

Rinfil and COMVOL conducted the instrumented steam-flow test of well 7 in 1970. The test revealed that the steam flow of the well was 27,000 to 32,000 lbs (12,150 to 14,440 kg) per hour. This meant that the well was capable of generating 540 kW of power using a noncondensing turbine, or 1,350 kW using a condensing type turbine.

By 1970, with the drilling of the seven diamond drill holes, six of which were exploratory and one production well, the preliminary research studies were completed. Geological, geochemical, and geophysical techniques had been used in the exploration of this first geothermal area and the results of all had been favorable. Meantime, some preliminary studies were being done on the other high-potential geothermal areas scattered throughout the country. No other exploratory drilling, however, was done.

The government agencies that were involved in geothermal energy exploration prior to the 1970s were mainly research agencies of the government. None of them was specifically charged to investigate or explore the potential of the country in geothermal energy. They did, however, do work related to geothermal energy data gathering and investigation in the course of their primary agency work. The two agencies that were most involved in direct investigation and exploration were the Bureau of Mines and COMVOL. The Bureau of Mines was involved because it was charged with studying the geology of the country relative to minable potential, and had the expertise and equipment in drilling. COMVOL was directly involved in studying the volcanic and solfataric areas of the country, and saw geothermal energy as an awesome gift of nature, and volcanoes as benefactors, not malefactors.

Identification

In November 1969, Ferdinand E. Marcos was reelected president of the Philippines. In his inaugural address he declared that the task of nation building was never ending. Consequently, the course the nation would take would be to address directly various problems, one of the most pressing being that of rural development.

The president was concerned with the balanced economic development of the country. Government attention was focused not only on the rapid industrialization of the urban areas but also on accompanying rural development.

Since the economy of the Philippines was, and still is, predominantly agricultural, there was a need to improve the production, processing, and distribution of the agricultural products. Irrigation facilities needed to be installed. Some degree of mechanization had to be introduced, especially in harvest and post-harvest activities. Intermediate raw material processing had to be introduced.

In all of these activities, power was a necessary factor. Infrastructure and communication facilities were planned to complement and support the rural development plans. The executive branch of government studied various ways of harnessing indigenous resources for maximum benefit. Many of the towns and barrios which needed a boost in development were located on islands. This compounded the problem of extending power and communication facilities to these areas. The challenge to the government was to seek technologies that were suitable to the needs of the country, and which made maximum use of local resources.

An official of the Philippine government, Executive Secretary Alejandro Melchor, in a trip to the United States in 1970 observed the operation of The Geysers in California and realized the potential of geothermal energy to suit local needs. Upon his return, he conferred with local experts and recommended to the president that commercial exploitation of the geothermal fields be initiated. He also recommended that, due to the preliminary research studies already conducted and showing favorable findings, the initial area developed should be the Tiwi, Albay geothermal area.

The president reviewed the recommendation and acted favorably on it by issuing, on August 1, 1970, Proclamation No. 739, "Establishing as Reservation for the Purpose of the Exploration, Development, Exploitation and Utilization of Geothermal Energy, Natural Gas and Octhane Gas a Parcel of land in the province of Albay, Island of Luzon, Philippines." The presidential proclamation set aside some 17,661 hectares (43,623 acres) in the geothermal area for exploration, exploitation, and development as a national reservation. The president had anticipated potential problems or conflicts over the ownership of surface rights in the area, and had paved the way for the more rapid development of the area. A previous law passed by Congress and signed by the president in 1967, RA 5092, had already provided that natural gases, methane, and geothermal energy belonged to the state.

Presidential Proclamation No. 739 identified Tiwi as the first area to benefit from the commercial exploitation of geothermal energy. This was particularly significant as the Bicol region, at that time, was one of the least electrified regions in the country.

Location

The Bicol region

The Bicol Peninsula, where the town of Tiwi, Albay is located, is on the southern tip of the island of Luzon. It is some three hundred kilometers (186 miles) south of Manila and is accessible by land, air, and sea. The Philippine National Railways system, as well as several bus systems, have regular daily trips to and from the area. There are four provinces in the Peninsula: Camarines Norte, Camarines Sur, Albay, Sorsogon; Albay is the leading province in economic terms. Total land area in the Bicol Peninsula is approximately 12,000 sq km (4,800 sq mi). Total population per the 1970 census for the Bicol region – which includes the islands of Catanduanes and Masbate – was 3.0 million.

Bicol is known for its volcanoes – especially the Mayon Volcano with its near-perfect cone – and its beaches. The climate is generally pleasant throughout the year. As it is within the typhoon belt, however, it is periodically visited by typhoons. The river basin areas of Camarines Sur and Albay are also subjected to flooding during the rainy season due to poor drainage.

The main economic activity in the area is agriculture with crops such as coconut, rice, and abaca (hemp) being dominant. The government of the Philippines considers the area to have a high growth potential especially for rice – once irrigation facilities are adequately provided – agribusiness, rural manufacturing and industry.

The Bicol region, unfortunately, has been described as among the most economically depressed in the country. In 1970, it had an unemployment rate 5 percent above the national average, and a per capita income rate 27 percent below the national average. A United Nations survey conducted at that time categorized Bicol as on a "downward transitional trend," meaning that a deteriorating economic situation existed. This was due to income maldistribution, a low level of physical and social infrastructure, the existence of tenancy problems in the agrarian sphere, and a high level of outmigration.

In terms of electricity, the Bicol region in 1970 was not connected to the main network of power stations, transmission, and distribution facilities of the National Power Corporation (NPC) in the Luzon area, known as the Luzon Grid.

There were only two major areas in Luzon not embraced by the Luzon Grid in 1970 – the Cagayan Valley in the north, and the Bicol area in the south. NPC, however, had two subgrids in the Bicol area, with a total of 5.2 MW in two small hydroelectric plants and a diesel plant. In 1968, only 20.89 percent of the households in the Philippines had electricity. In the Bicol region, however, the figure was 7.39 percent.

The government had plans for the overall development of the region in the early 1970s. For this rural development scheme, power was a critical ingredient. Therefore, the commercial exploitation of the Tiwi Geothermal Field assumed added significance.

In 1973, President Marcos issued Executive Order 412, creating the Bicol River Development Program, an integrated area development program of the national government. The main development strategy of this program was to build up the physical infrastructure, improve land tenure arrangements, increase agricultural and labor productivity, and encourage private investment in agribusiness and other rural-based industries. The end result expected by the 1980s was an improvement in the quality of life of people in the program area and the Bicol region in general.

Tiwi

The municipality of Tiwi is situated in Albay Province, Bicol. It has an area covering 12,342 hectares (30,455 acres), or 5 percent that of Albay. In terms of population distribution, the town is twelfth in size in the province, with a population of a little more than 20,000 people. The town is located 40 km (25 mi) north of the capital city of Legaspi and some 530 km (329 mi) from Manila. The most important occupation in the town in 1970 was farming, followed by fishing and cottage industries. Although it was located, economically speaking, in the most important province, Bicol, the municipality of Tiwi in 1970 was not electrified.

The geothermal resource in the town has been put to use in various ways. Food is sometimes cooked over the boiling mineral water springs – chicken is dressed and steamed there, and eggs are boiled. Copra and palay grain are dried over the hot igneous rocks. Thermal spring resorts also existed in 1970.

Formulation

Late in 1970, Executive Secretary Melchor, who was also concurrently chairman of the Power Development Council, the highest coordinating and policy-making agency in the electrification program of the country, started initial negotiations with Union Oil of California, the owner of The Geysers, to develop local geothermal resources. Union Oil officials signified their willingness to participate in the geothermal program of the Philippines.

The executive secretary then constituted a panel to discuss with Union Oil officials, the nature of possible collaboration. Named to the panel were: the general manager of the National Power Corporation, the main government agency charged with power development, as chairman; and representatives of the Power Development Council, COMVOL, the Bureau of Mines, and the Executive Office as members. Strong government support was given to the panel, which was empowered to call upon any government department, bureau, office, agency, or instrumentality for assistance in the pursuit of the project. The executive secretary, who in effect carried authority emanating from the president's office, facilitated the coordination work.

Union Oil officials, led by the head of their geothermal division, visited the Tiwi geothermal area and reviewed some of the research findings. Based on their observations, they agreed that the area showed great promise for geothermal exploration and development.

An initial agreement made with Union Oil was that they, and their local subsidiary, Philippine Geothermal Incorporated, would be service contractors for handling exploration and extraction of steam. The main government agency supervising the project and running the necessary power generating facility would be the National Power Corporation.

Service Contract with PGI

Since the country had yet to acquire expertise in the commercial exploration, exploitation, and development of geothermal resources, a service contract was signed between the National Power Corporation through the office of the economic coordinator, and Philippine Geothermal, Inc. (PGI) on September 10, 1971.

Philippine Geothermal, Inc. was the Philippine subsidiary of Union Oil of California, a company that was then ranked fifth in international oil drilling and first in geothermal exploitation for power generation in the United States.

The committee, chaired by the general manager of the National Power Corporation, worked for the immediate commercial development of the geothermal reservation for electric power generation. PGI, once it was convinced of the geothermal potential of the field, was interested in pursuing the contract at an economically acceptable return.

The service contract effected can be described as a "fuel supply contract." PGI would be the fuel supplier, providing the steam to be used by NPC in power generation. PGI would also take responsibility for the disposal of the effluent. Provisions of the contract stipulated that:

> NPC shall provide Foreign Exchange and Philippine Currency in the amount required to pay 55% of the total operating cost incurred by PGI on behalf of NPC in conducting Exploitation and Effluent Handling Operations. (Subsection 5.2) As reimbursement for PGI's operating costs, NPC shall pay PGI 4.55 U.S. mills per kWh of electric energy generated until the operating cost will be fully reimbursed. (Subsection 7.2(a)) In addition to the payments under (a) above, NPC shall pay PGI 5.45 U.S. mills per kWh of electric energy generated. . . .for service rendered to NPC pursuant to this contract. (Subsection 7.2(b))(1)

Under the terms of this contract, then, NPC would only pay PGI for a major part of its expenses when the electric power was generated. PGI was committed to explore the field and inform the NPC when it had discovered geothermal energy in sufficient quantity to supply a generating plant with a capacity of at least 55,000 kW. PGI was also to provide the NPC with technical assistance on the location and design of plants and equipment for the conversion of geothermal steam into electric power. NPC, on the other hand, was to handle the generation and distribution of electric power.

The contract period was for 25 years, renewable for another 25 years. Since the contract has been in effect, however, some changes have been made with the approval of both parties. The terms as quoted are based on the third amendment made on July 1975 and have since been revised. The original agreement pegged the cost of steam generation to the price of oil. This became unrealistic after the oil crisis in 1973. Revisions pegged costs instead to more objective factors, and provided that NPC's share of the total expense in dollars and pesos would not exceed 55 percent.

Feasibility Study and Appraisal

In 1973, the National Power Corporation contracted Rogers International to prepare a project feasibility study on the establishment of a geothermal power plant in Tiwi, Albay. The immediate users of the power to be generated would be the service area of Camarines Sur, Albay and Sorsogon, which in 1973 were electrically isolated from the Luzon Grid. The eventual plan of the NPC was to include the Tiwi power output as part of its overall expansion program for the Luzon Grid. By 1976, the lines of the Luzon Grid would be expected to meet with those at the Tiwi area completing the Luzon Grid expansion program.

The feasibility study was finalized in April 1973. Per the schedule given in the study, final design work was to be started as early as May of the same year, for the project to be completed by June 1975.

The introduction to the study clearly recognized the role of the NPC as the government agency charged with the development of power for electrification as well as for other uses. In support of this, it cited the following: Republic Act No. 6038 declared it a policy of the state to pursue and foster the total electrification of the Philippines on an area coverage basis as being vital to the welfare of the people and the sound development of the nation. Republic Act No. 6395 declared that 1) the comprehensive development, utilization, and conservation of Philippine water resources for all beneficial uses, including power generation, and 2) the total electrification of the Philippines through the development of power from all sources to meet the needs of industrial development and dispersal and the needs of rural electrification, are primary objectives of the nation.

The introduction further discussed the status of power development and NPC programs for the coming years. The capacity of NPC generating stations servicing the Luzon Grid in 1972 was approximately 500 MW. The NPC had a transmission line network totaling 2,725 km (169 mi) of lines, and covering the greater part of Luzon. The Luzon Grid directly or indirectly served 281 cities and municipalities.

NPC had embarked in 1973 on a program to electrify the rural areas and promote development throughout the country in line with government objectives. Hence, an expansion program which included the increase in power-generating capacity and extension of transmission lines was started and funded by a fifth power loan from the International Bank for Reconstruction and Development (IBRD) of the World Bank. The grid was expected to reach Camarines Sur in the Bicol Peninsula by mid-1976.

Findings indicated that the feasibility study was designed to provide 21 MW of electricity to the service area of Camarines Sur, Albay, and Sorsogon. Area population of the three provinces according to the 1970 census was slightly over two million. A total of 50 towns was expected to be serviced by the power generating unit. At the time of the study, 13 towns of this area were being served by other power stations of the NPC, 21 had power at some time of day from self-generating units, 16 had no power whatsoever.

An analysis of the market indicated that the total potential market for the Tiwi power far exceeded the 21-MW capacity of the pilot plant. This potential market would have to be developed with active support from the National Electrification Administration, a government agency charged with the development of electric cooperatives to operate rural power distribution systems – from the main stations of the NPC to the final end users. For its first year of operations, therefore, the power utility was expected to operate at a relatively low rate of 24 percent of capacity. For subsequent years, the output would be 90 percent of capacity, since the service area demand was expected to grow, and the transmission lines to the Luzon Grid would be connected in 1976.

The supply of the steam for the geothermal power plant would come from the PGI wells. PGI had so far drilled two production wells and were in the process of drilling the third. Two additional wells were needed to ensure the steam supply requirement for the 21-MW plant. In total, 13 wells were necessary to maintain the steam supply for 35 years, the expected life of the geothermal power plant.

The estimated total cost of the project was ₱14,687,000, broken down as follows:

Direct Cost	Pesos (in thousands)
A. Production plant	₱5,009
B. Transmission plant	163
C. Transmission lines and substations	5,825
Indirect Costs	
Engineering, contingency, and interest	3,690
Total	₱14,687

Rogers International Consultants considered that building power plants of this size would enable the operators and the management staff to gain experience with the generation unit and field situations inherent in geothermal power generation. Future unit sizes, they felt, could be much bigger to correspond to the growth of the demand for electricity. It was also noted, however, that building a small plant such as the one planned was relatively expensive – meaning that the power plant would not enjoy favorable economies of scale. A 50-MW size plant would be more favorable from the point of view of economics, and would result in a 25 percent decrease in electricity costs per unit of power.

The design for the project included the setting up of two geothermal power units of 10.5 MW each in Tiwi, Albay, and 317 km (196 mi) of 60-kW transmission lines from the Tiwi plant to various points in the three provinces.

Philippine Geothermal, Inc., as the supplier of the steam, initiated the drilling of production wells to provide the necessary steam. As of the time of feasibility study preparation, PGI had drilled two wells, Naglagbong one and two, and was in the process of drilling a third. The first two wells were commercially favorable, with hot water produced at a reservoir temperature of 520°F (256°C) to 535°F (265°C). The first well produced at an initial rate of 350,000 lb/hr (157,500 kg/hr) mass flow; the second well at a rate double that of the first. The three wells proved the existence of a reservoir area of 800 acres (320 hectares). An additional two wells would be drilled to complete the five wells needed to supply steam to a 21-MW power plant, assuming that a mass flow of 500,000 lb/hr (225,000 kg/hr) from each well were realized.

For the initial development of the 21-MW capacity, a spacing of 40 acres per well was planned so that the 13 wells would produce from an area of 520 acres (208 hectares). The geologic setting and reservoir rock at Tiwi are similar to that at Wairakei, and the reservoir fluids are at about the same temperature. In Tiwi, the generating potential based on proven acres, therefore, was expected to closely parallel that of Wairakei. If so, the 800 proven acres (320 hectares) would have a generating potential of 430 MW and the total 5,600 acres (2,240 hectares) a potential of 4,000 MW. If a spacing of 40 acres (16 hectares) per well, or even 20 acres (8 hectares) per well were maintained, this would result in a producing capacity of 700 to 1,400 MW from the potential of 5,600 acres (2,240 hectares).

The geothermal production was a mixture of steam and water. PGI was to construct a gathering system (see fig. 4.2) to separate the steam from the water, and transport the steam to the plant. The residual water and any cooling tower effluent would be gathered, cooled in open but protected gutters, and eventually emptied off into the ocean. Since the water from the Tiwi springs contained minimal chemicals – dissolved sodium chloride primarily – and the noncondensable gases consisted primarily of 99.9 percent carbon dioxide (trace amounts of methane, ethane, propane, hydrogen, and nitrogen were found; no ammonia or hydrogen sulfide), pollution control considerations were concentrated on the cooling of the runoff water, which would be by gutters. The steam would be delivered to the power plant by PGI in the

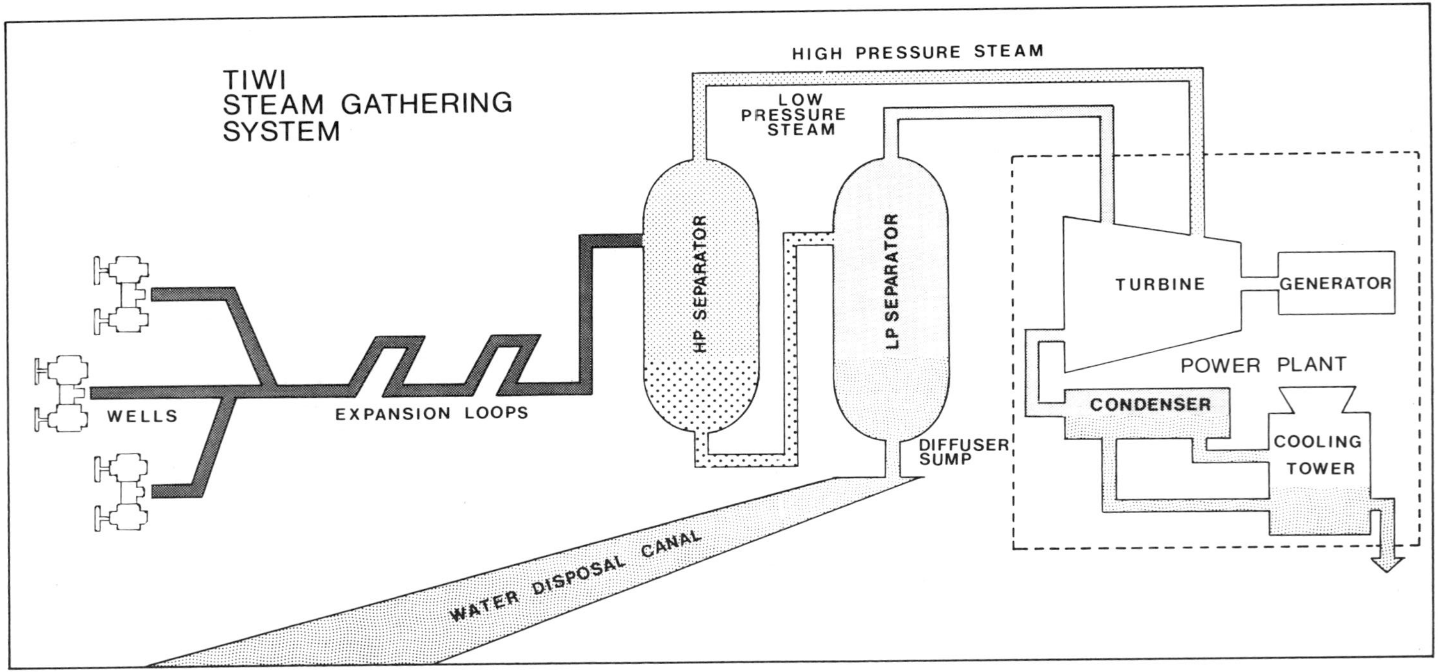

Fig. 4.2. Tiwi steam-gathering system.
Source: National Power Corporation, Manila, Philippines.

following manner: flash steam from the geothermal field would be delivered to the power plant at the designed turbine throttle pressure of 95 psig. Approximately 84 percent of this plant steam will be expanded through each of 2-by-11,000 kW (gross) turbine-generator sets and the steam exhausted to the condensers operating at 4 inches (10 centimeters) mercury vacuum. The remaining 16 percent of the steam will be used as power fluid for the steam jet gas ejectors.

To permit starting the plant without an external power source, a 500 kW cold start turbine was to be provided and made common to the two units. The electrical generation system was to consist of two separately driven air/water cooled 11.0-MW gross generators, each separately connected through its own main power transformers to the 60 kV outdoor bus.

The general facilities would consist primarily of the turbine generator building, a reinforced concrete building. At the geothermal plant, a substation rated at 24 kVa, 13.8/69 kV, with three feeders, would be built for take-off to the service area.

National Power Corporation's share of the initial investment for developing the 21-MW production capacity was US$452,000. Total costs were expected to amount to 4.81 mills/kWh computed at an average load of 34 percent for the first period of production and 90 percent after January, 1977 when the 130-kV line was to be completed.

Expansion of the field at the rate of 50 MW per year starting mid-1977 would require an NPC initial investment in wells and facilities in the United States equivalent of $1,081,250. Operating for the 35-year life of the plant, each 50-MW expansion would cost 4.11 mills/kWh computed at an average load of 90 percent.

The time table for completing the plant was worked out using two scheduling techniques, the Critical Path Method (CPM) and Planning, Evaluation and Review Techniques (PERT). These methods indicated that it would take two years to complete the power plant and facilities, only if special measures were taken to permit early commitment of equipment. The plant would be located in such a way that it would be central to the production wells to minimize the cost of the steam-gathering system. In the study, the topography of several sites for the power plant was discussed and a particular site was recommended.

Appraisal

NPC requested Rogers International to recommend an independent consultant to evaluate the findings of PGI on the geothermal potential of Tiwi.

Professor H.J. Ramey, Jr. of Stanford University was therefore engaged in February 1973 to review the PGI reports.

Professor Ramey reviewed the findings of PGI on the extent and main characteristics of the Tiwi Geothermal Field Reservoir. Based on the results of the various tests, volumetric calculation, and resistivity surveys done by PGI, he concurred that there was likely to be sufficient steam reserves for a 21-MW pilot power plant at Tiwi. He therefore enthusiastically endorsed the project in a letter written to Rogers International and dated April 1, 1973. Unfortunately, however, Professor Ramey was not able actually to visit the project in the Philippines. His evaluation was based on data furnished him by Union Oil.

Despite the findings of the feasibility study showing that it was technically feasible to operate a 21-MW power-generating plant, the NPC general manager decided not to implement the project immediately. Instead, further studies were conducted by PGI on the field. This continued exploration work yielded favorable results, and indicated that a bigger generating capacity plant could be supported. This would result in better economies of scale and therefore cheaper per kilowatt costs, as discussed in the feasibility study.

In 1976, the engineering department of the NPC revised its design for a larger-capacity power plant. No further planning studies – such as another feasibility study – were carried out. Since the NPC had been involved in the construction and operation of power plants since 1930, no major difficulties were anticipated. The government agency had qualified personnel to design and supervise the job.

PHASE 2: SELECTION, APPROVAL, AND ACTIVATION

Selection and Approval

The project was approved in principle by the National Power Corporation and the Power Development Council from the time the service contract was signed with PGI in 1971. Since it had the strong backing of the executive office, approval of the power plant was confirmed. The National Power Corporation, as the government agency empowered to develop and produce electric power demands of the country, was the implementation government agency for the project.

The corporation had received a stronger mandate from the president of the Philippines, who issued Presidential Decree No. 40 in 1972 restating the policy of the government to effect the total electrification of the country and indicating the role of the National Power Corporation in this task: "...The setting up of transmission line grids and the construction of associated generation facilities in Luzon, Mindanao and major islands of the country, including the Visayas, shall be the responsibility of the National Power Corporation (NPC) as the authorized implementing agency of the state."(2)

By 1973, the Bicol River Development Program had also been initiated under Executive Order 412 (later amended by Presidential Decree 926). This integrated area development scheme relied on the development of the Tiwi Geothermal Field as a major source of power. The rural development of the area was linked with the Tiwi Power Project; hence, the project was strongly endorsed and approved.

Activation

Power plants

Under the revised development scheme, two power generating units of 55 MW each were to be installed initially at Tiwi. Because these were of a larger size and would also require a larger supply of steam, the power plants were only expected to be operational in the 1978-79 time period.

A contract for these two units was signed by NPC and Mitsui Company, Limited of Japan after an international bidding in which the Mitsui bid was the most acceptable. The contract was signed on December 21, 1974 and included the design, manufacture, and delivery, (CIF) cost, insurance, and freight at Tobaco Port in the province of Albay, of major electromechanical equipment and machinery, and the erection of a cooling tower. The total amount of the bid was 5,872,703.00 Japanese yen, or approximately US$29.35 million.

Another contract was negotiated with Mitsui Company, Limited and signed on September 6, 1976. This contract called for the design, manufacture, and delivery of the balance of the plant, testing and supervision of preliminary operation. The contract amounted to 4,520,981.00 Japanese yen, or approximately US$22.61 million.

Civil works

Tenders were also called for the construction of the powerhouse facilities. The bid was awarded to a local contractor, F.F. Cruz and Co., in the form of a lump sum contract in the amount of ₱15,775,997.15. The effective date of the contract was August 27, 1976; contract duration was for 612 calendar days.

Civil works also included the construction of a housing complex for Tiwi field personnel. Again a local contractor won the bid for the housing complex at ₱12 million.

PHASE 3: OPERATION, CONTROL, AND HANDOVER

The National Power Corporation

The NPC is a corporation wholly owned by the Philippine government. It was created under Commonwealth Act No. 120 on November 3, 1936 with the primary objective of undertaking the development of hydroelectric and other natural sources of power generation. In 1960, under Republic Act No. 2641, the NPC was converted into a stock corporation, wholly government owned, with a capitalization of ₱100 million. The authorized stock was increased to ₱300 million under Republic Act No. 4897 in 1967.

In 1971, Congress approved Republic Act No. 6395, titled "An Act Revising the Charter of the National Power Corporation." Guiding the policies of the NPC was the provision of the act which states:

> Declaration of Policy – Congress hereby declares that 1) the comprehensive development, utilization and conservation of Philippine water resources for all beneficial uses, including power generation, and 2) the total electrification of the Philippines through the development of power from all sources to meet needs of industrial development and dispersal and the needs of rural electrification are primary objectives of the nation and shall be pursued coordinately and supported by all instrumentalities and agencies of the government, including its financial institutions [emphasis added].(3)

Since the energy crisis in 1973, the government dedication to power generation and the development of indigenous energy resources has been accelerated. Essentially, the electrical power development program, as enunciated by the state, is aimed at attaining self-reliance through the availability of the nation's energy resources. Presidential Decree 40 reserved exclusively to NPC the development and exploration of the country's power industry. This has meant that the National Power Corporation is one of the energy companies falling under the purview of the Ministry of Energy. In 1979, the capitalization of the NPC was increased to one billion.

In 1973, the National Power Corporation had seven main offices or departments: engineering and construction, administration, special projects, finance, and the regional offices: Luzon regional office, Visayas regional office, and Mindanao regional office (see fig. 4.3).

The planning of projects was the responsibility of the engineering and construction office. Projects were implemented by the regional office where the project was located.

For the Tiwi project, the engineering section was responsible for project development studies, engineering services (which included pre-engineering studies such as geology and geotectonics), site exploration works, plans and feasibility studies, designing studies, and the preparation of specifications and tender documents.

A section of the administration department looked after the preparation of contract documents, the screening of bids, and the awarding of contracts.

The Luzon regional office took responsibility for the actual field project implementation of the Tiwi project. This meant that the project manager at the site would report to the Luzon regional manager.

In 1974, a geothermal division was set up at the NPC to supervise the geothermal exploration and exploitation work. The division was under the engineering and construction department. Its main concerns were to review work done by the PGI to ensure the supply of steam for the projected power plants, to protect the interests of the Philippine government in matters concerning dealings with the service contractors, and to embark on exploration and developmental activities in other promising geothermal areas. The COMVOL chief volcanologist, Alcarez, was transferred to the NPC to head this division.

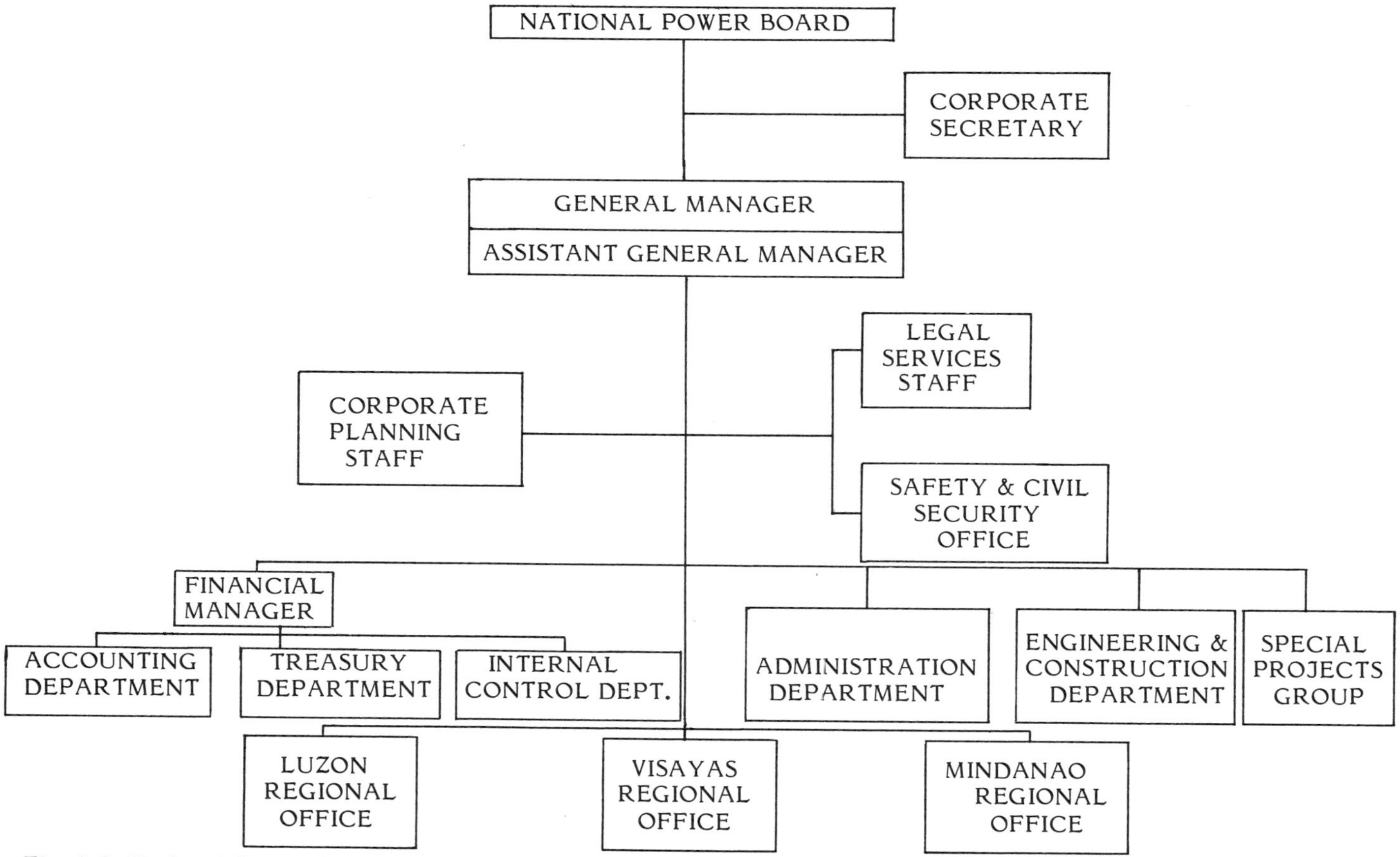

Fig. 4.3. National Power Corporation – organization chart as of December 1977.

The organizational chart for the geothermal division in 1974 is shown in figure 4.4. For the duration of the project, PGI maintained both a head office in Makati, Metro-Manila, and a field office at Tiwi.

A steering committee was created as set forth in the service contract to prepare policies and operational plans, and to coordinate all the activities related to the geothermal project from initial steam exploration to final generation and distribution of power. The committee was composed of 12 members: six from NPC and six from PGI; the chairman was an executive vice-president of NPC.

The committee would review, at the start of a year, a proposed work program and budget presented by PGI for the exploration and exploitation of geothermal resources. The annual work program would be submitted at least three months prior to the beginning of a fiscal year. Based on the committee's approval, the work program would be put into effect, and would be the basis for evaluating progress during the year. NPC was given the right to propose revisions in PGI's work program.

For the duration of the project, the NPC had to coordinate with three main contractors: PGI, F.F. Cruz, and Mitsui. Consultants – namely the Tokyo Electric Company (TEPSCP) – were contracted on the structural and architectural work.

The project was implemented smoothly with no major problems occurring.

Under the development program, the actual construction of the power plant was to start in 1976, while the exploration work was to have been underway all along. By December 1978, the first 55-MW power unit was expected to be operational.

By March 1978, the development reports on the geothermal field indicated the following information: twenty-seven holes had so far been drilled to an average depth of 1,810.85 m (5,976 feet). The type of geothermal system developed was liquid dominated (well production was steam and hot water). The average mass flow of the wells was 557,087 lb/hr; the average steam flow was 122,020.7 lb/hr. The average steam:water ratio was 21.76 percent.

Because the average steam flow was 122,020.7 lb/hr, it was expected that 20-odd production wells would be needed to generate the initial capacity of 110 MW. Reservoir fluid contained between 5,000 and 6,000 ppm total dissolved solids, mostly chlorides of calcium, silica, and potassium.

A satellite system type of gathering system was implemented for the Tiwi area, since it was a liquid dominated field. Under this system a mixture of steam and hot water from a cluster of producable wells is gathered in a substation where high pressure steam and water are separated. As the high-pressure water still contains considerable heat, this water is then flashed to a lower-pressure separator. In the central plant station, the water is again flashed to a secondary pressure separator. The steam enters the turbine through two separate nozzles at different pressures. Separated water from the central plant station goes to the sump pit and is then disposed of in the ocean via protected gutter or canal water disposal system.

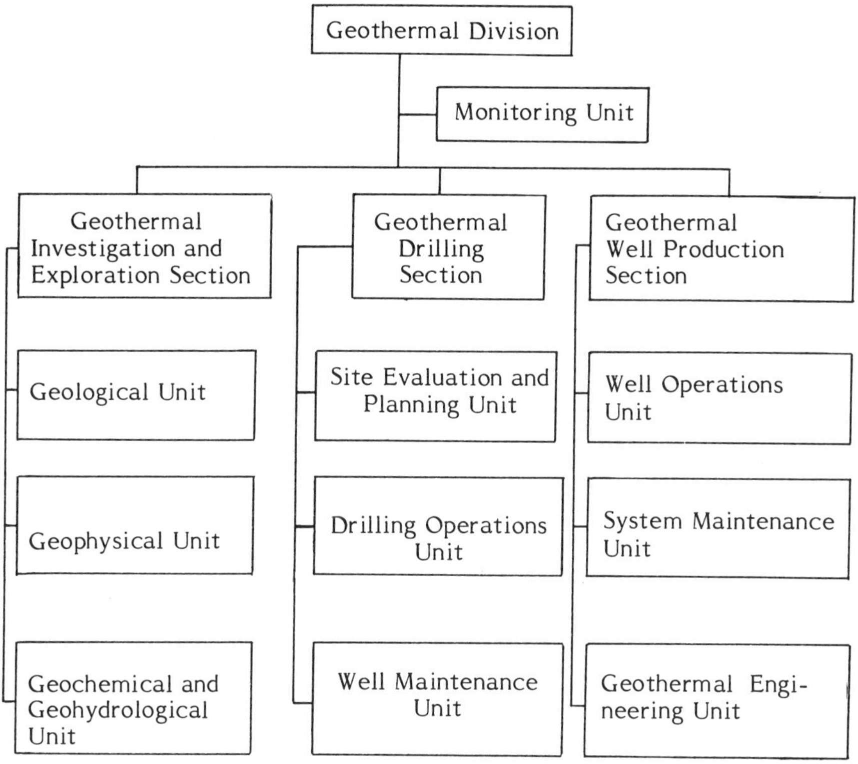

Fig. 4.4. Geothermal division of National Power Corporation – organization chart, 1974.

The construction of the power plants suffered only minor delays. During the construction of powerhouse 2, isolated soft clay materials were encountered at the depth of the turbogenerator foundation. Further boring tests were conducted, resulting in an estimated seven-month delay. The NPC was able to coordinate with the contractor and the activities were accelerated, so the net delay was only four months.

Completion and Handover

The first geothermal power plant was inaugurated in January 1979 by President Ferdinand E. Marcos. The second was inaugurated in June 1979. The president, in his message, highlighted the significance of the geothermal program and the government's policy of development and self-sufficiency. It was due to the country's increasing need for energy, the President stated, that he extended full support to indigenous energy projects such as the geothermal project at Tiwi.

By the time Tiwi Power Plant 2 was inaugurated, the Philippines had accelerated the geothermal program. A 55-MW unit was also inaugurated in the Makiling-Banahaw area, and a 3-MW unit in Tongonan in 1979. This meant a total of 168 MW of electricity was generated from the geothermal energy resources. The Philippines thus became the world's fourth leading producer of electric power from geothermal sources and may soon become the major producer of geothermal energy from hot water fields in the entire world.

The approximate project cost of the Tiwi Geothermal Power Projects 1 and 2 totaled 445.48 million, with ₱389.76 representing the foreign portion and the balance the local portion.

Output of one 55-MW power plant at Tiwi is equivalent to a crude production of 45,000 barrels of crude oil per year, according to geothermal economists. This means a savings in foreign exchange to the Philippines of $8.64 million annually at a price of $16.00 per barrel. The economic benefits of the Tiwi Geothermal Power Plants are therefore immediate and tremendous.

Environment Analysis

Chemical analysis of the thermal waters and steam at the Tiwi geothermal field indicates a minimal effect on the environment and ecosystem.

The main environmental harm can come from three sources:

1. The heat of the waste water if it is to be fed into a nearby river or stream,
2. The chemical elements that may be present in the waste water, and
3. The chemical elements that are gases released into the atmosphere.

Because of the high steam purity, only a small percentage of the geothermal energy is in the form of hot water. This waste water, however, must be disposed of in an acceptable form and temperature. At present, the water is cooled through a system of open ditches or canals before it reaches the Philippine Sea.

Samples taken of the water are periodically analyzed and the dissolved solids have been found to be primarily sodium chloride. This does not pose a problem since the waste water is disposed of in the Philippine Sea, a saline body of water that opens out into the ocean. Total dissolved solids in each sample are approximately 10,000 ppm.

The steam from the wells has been sampled to determine the composition and percentage of the noncondensable gases present. The gas was found to consist mainly of carbon dioxide (99.9 percent) with trace amounts of methane, ethane, propane, hydrogen, and nitrogen. Two-stage ejectors are provided on each turbine exhaust steam condenser to remove the noncondensable gases entering with the steam. These gases are vented directly to the atmosphere through a vent silencer.

The main responsibility for pollution control in the Tiwi project is with PGI, whose contract with NPC included the gathering system for the steam supply, steam separators, and the disposal of the effluent. When the Ministry of Energy was created in the mid-1970s, a separate body, the environmental unit, was set up. Both PGI and the environmental unit are studying the economics of reinjecting the hot water into the earth through the dry wells.

Other uses for the excess steam and hot water are under study. At present, a pilot salt-making plant exists in Tiwi, part of an experiment to develop geothermal salt in commercial quantities. Experts from three government agencies – COMVOL, the National Science Development Board, and the Philippine Navy – are involved in this activity.

Another team is working to utilize steam in the drying of copra and palay. This will greatly benefit the area, as the Bicol region does not have a pronounced dry season and there is a year-round abundance of tropical rainfall.

PHASE 4: EVALUATION AND REFINEMENT

Evaluation and Policy Statements

The Tiwi Geothermal Project, a project of the Philippine government, was conceived as a means to hasten the development of rural areas in the country through the development of an indigenous energy resource.

The Tiwi project involved a number of government agencies from the time of its conception. It was planned and implemented just prior to the time when the energy situation in the world was undergoing drastic changes. In consequence, numerous changes in organization for energy development occurred. What emerged eventually in the mid-1970s was a comprehensive government energy program under a newly created Ministry of Energy. In the words of Energy Minister Geronimo Z. Velasco, in a message included in the publication of the "Ten-Year Energy Development Program, 1978-1987":

> Our new theme is ENERGY FOR DEVELOPMENT. The Ministry of Energy, having studied the magnitude and complexity of the energy problem, has designed a cohesive and coherent national energy policy for the future. The Ten-Year Energy Program. . .is the result of all our efforts. It is a modest attempt by the Ministry of Energy to integrate all related activities in the energy sector into a single, coordinated energy plan. Its scope is far-reaching. It spans a whole range of activities, from energy-resource development through power transformation and rural electrification to the downstream activities that ensure country-wide fuel distribution in forms usable to the ultimate consumers. The program will affect the nation for generations to come.(4)

The Tiwi case has highlighted several important issues and observations:

1. Coordination of energy-related government agencies is important. A number of government entities played a significant role in the Tiwi project. The relationship of these government agencies and their role in the overall energy program had been changing. The initial agencies involved in the project were operating mostly independently, and at times with committees or superstructures, to coordinate the work. By the time the project was completed with the inauguration of the first and second geothermal power plants at Tiwi, a centralized Ministry of Energy had been created to integrate all energy programs.

2. There is a need to develop indigenous energy resources. In the Philippines, the need to develop indigenous energy resources was recognized before the oil crisis. This was significant in the light of the fact that the geothermal fields were explored and developed and producing electricity during the 1970s. Geothermal fields that have been explored after the oil crisis may only produce power by the 1980s. This circumstance would imply two things: a longer lead time for exploration and development when alternative energy is badly needed, and more expense in terms of development of the wells, since this requires oil fuel, the price of which is rapidly escalating.

3. Top administrative support for the project ensured the success of the project. Because of the support of President Marcos, and through him, the assistance of Executive Secretary Melchor, the project cut through red tape and was able to proceed at a relatively efficient rate. In line with his policy of developing local energy sources, the president has set aside, as government reservations, areas of commercially exploitable geothermal fields – among them Tiwi.

4. Manpower development in the energy field has been constant. Philippine scientists and professionals have been trained constantly both locally and abroad. The government has made use of expertise from the countries in the world that have commercially exploited geothermal energy – the United States, New Zealand, and Italy. Whether consultants are brought into the country or local scientists go for studies and training in a foreign country, the process of technology transfer has been constant. Whereas at the start of its geothermal development, the country had to rely on service contractors to develop the fields, Philippine service contractors, since 1977, through the Energy Development Corporation, have taken over the commercial development of the geothermal fields, such as the one in Tongonan, Leyte.

5. Leadership in early geothermal development was a vital element. In the Philippines, it was provided by the chief volcanologist of the COMVOL, Arturo Alcaraz. He has been referred to in the country as the "Dean of Geothermal Energy." Much of the perseverance in geothermal energy studies, much of the training and development of geothermal expertise in the country has been ascribed to him. A quick review of experts in geothermal energy in the country – at the NPC, Ministry of Energy, COMVOL, and even at PGI – indicates that many geologists, engineers, and other scientists have trained with Alcaraz.

6. Administration by the government agency was effective. The National Power Corporation, in assuming responsibility for the com-

mercial development of the geothermal field, has proven, not for the first time, its ability to orchestrate the tremendous job of constructing a power plant. The operations of the NPC are scattered all over the country and cover hydroelectric, nuclear, and geothermal projects, and installation of transmission lines. The NPC, too, has been dynamic in its organization and management. Recently, a reorganization was effected wherein some projects are now handled on a purely project management basis.

7. Continual evaluation of project effectiveness is necessary. The NPC and the Ministry of Energy have conducted an ongoing evaluation of the Tiwi geothermal experience. Some of the effects of the evaluation have been immediately implemented. These include revisions in the service contract with PGI and the drafting of new service contract arrangements for future projects, and the evaluation of the contractors and their performance in the light of possible future contracts. The Tiwi experience in many ways is a pilot project, the NPC's first experience in the commercial exploitation of geothermal energy.

No comprehensive post-project evaluation has been conducted to this date. However, future evaluation studies of this and other energy projects are planned by the Ministry of Energy.

The 1978 Philippine Energy Program

In 1977, the Philippines was 95 percent dependent on crude oil for its commercial energy requirements. Since the country did not produce oil in commercial quantities, this meant that all the crude oil was imported.

Table 4.2 and figure 4.3 indicate the country's dependence on oil and the effect on the national balance of trade.

Due to the urgency of the situation, President Marcos launched programs for extensive energy development and energy conservation. Efforts were stepped up to develop and tap local energy resources for energy self-sufficiency.

On October 6, 1977, a Department of Energy – now a Ministry of Energy – was created in recognition of the vital importance of energy to the national well-being, as well as of the multidimensional facets of energy development and use.

The Ministry of Energy has prepared comprehensive plans and programs for energy development and use. In 1978, it published the first "Ten-Year Energy Development Program, 1978-1987." Since then, it has identified innovative projects and programs, as well as modified or strengthened existing ones to suit the dynamic needs of the developing country.

An assessment of indigenous energy resources made by the ministry in 1979 indicates that of these resources, only coal, hydro, and geothermal resources have been commercially developed. The potential for development of the following indigenous resources is shown in table 4.3.

Table 4.2. Oil Imports and the National Balance of Trade (CIF in Millions US$)

	1973	1974	1975	1976
Oil imports	231	681	833	917
Total imports	1,790	3,468	3,776	3,953*
Oil imports as a percentage of total imports	12.9	19.6	22.1	23.2
Total exports	2,099	2,974	2,610	2,964*
Oil imports as a percentage of total exports	11.0	22.9	31.9	30.9
Balance of trade	309	(494)	(1,166)	(989)
Oil imports as a percentage of negative balance of trade	–	138	71	93

*Estimate

Source: Central Bank, as quoted in "Ten-Year Energy Development Program 1978-1987." (Makati, Metro Manila: Ministry of Energy, 1978), p. 17.

Within the next ten years, the government's target is to reduce drastically the country's dependence on imported oil. In 1988, it is expected, about 54 percent of the country's energy requirements will be met by local sources of petroleum, hydroelectric, geothermal, coal, solar, nuclear, and nonconventional energy sources. This means a reduction of 41 percent from the 95 percent dependence on crude oil in 1977. The government, therefore, will embark on more energy development projects, as well as highlighting the need for energy conservation in the country.

An organizational chart of the Ministry of Energy, showing some of the public corporations involved in energy programs, is shown in figure 4.7. (The Power Development Council, a policy-making body, was abolished with the creation of the Ministry of Energy.)

The strategy for development and attainment of the national energy objectives is spelled out in the energy program. Yearly targets are broken down into targets for the different energy resources. Projects are then launched and monitored to ensure that the targets are obtained. For the geothermal sector, for example, the program is presented in table 4.4 (see also fig. 4.8).

The expansion program has produced immediate and impressive results. In 1979, the first large-scale geothermal power units and first commercial production of indigenous petroleum were inaugurated. As a result, 900 MW in power-generating capacity were added to the national system.

The yearly energy output goals, classified into type of energy resource, has made the entire energy program manageable. The energy

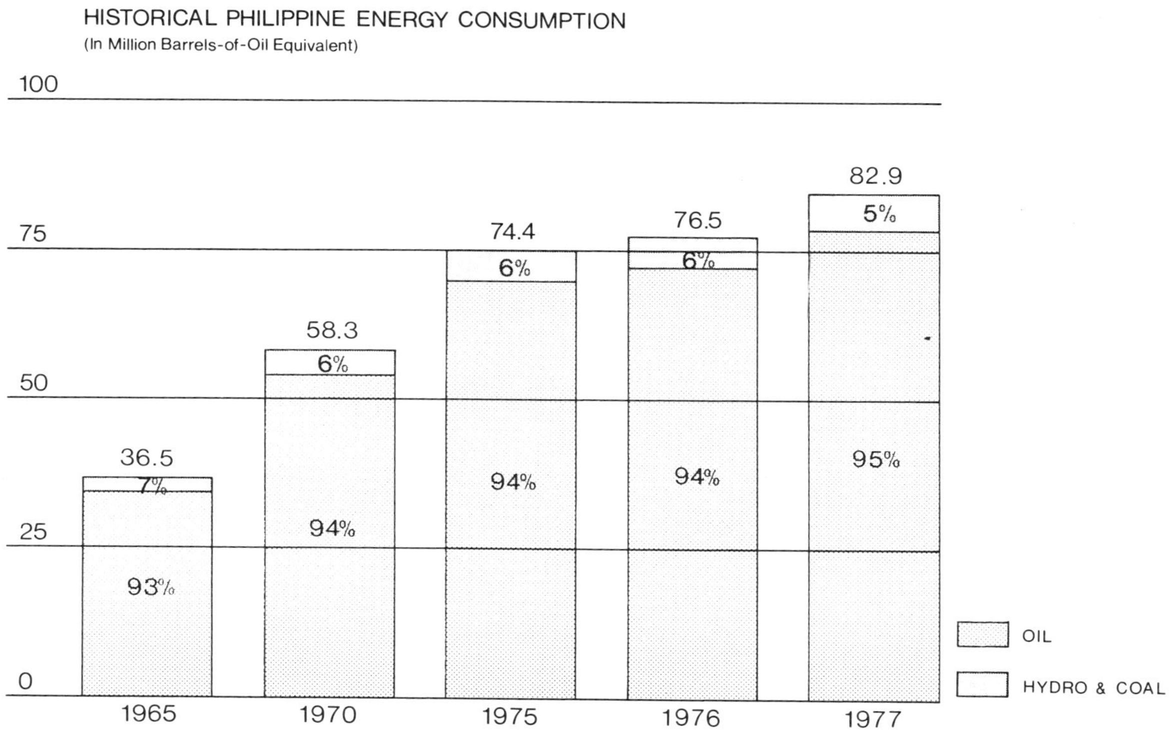

Fig. 4.5. Historical Philippine energy consumption (in million barrels-of-oil equivalent).
Source: "Ten-Year Energy Development Program, 1978-1987" (Makati, Metro Manila: Ministry of Energy, 1978), p. 17.

Table 4.3. Tentative Assessment of Indigenous Energy Resources (in Million Barrels of Oil/Oil Equivalent[MMBOE])

	Developed	Probable	Potential[a]
Finite			
Petroleum	Under evaluation	Under evaluation	Under evaluation
Coal	220	180	4,000
Uranium	–	10	Under evaluation
Regenerative			
Hydro	4.4/year	36/year	50/year
Geothermal	3.3/year	26/year[b]	2.30×10^3/year
Nonconventional (solar, biomass, etc.)	0.45/year[c]	16/year[d]	1.43×10^3/year[e]

[a]Including developed probable.
[b]In the fields currently under development.
[c]Usage directly resulting from government programs only.
[d]Estimated maximum usable for 1979.
[e]Assuming 10 percent land area usage, 5 percent average efficiency.

Source: "Ten-Year Energy Development Program 1979-1988" (Makati, Metro Manila: Ministry of Energy, 1979), p. 11.

Table 4.4. Geothermal Development Program

Year	Target Number of Fields	Number of Wells	Cumulative Geothermal Steam Availability (MW)	Cumulative Installed Generating Capacity (MW)
1978 (actual)	3	30	305	3.0
1979	4	59	470	278.0
1980	4	68	635	445.5
1981	6	48	740	561.5
1982	7	48	1,015	561.5
1983	7	48	1,125	561.5
1984	8	41	1,290	561.5
1985	8	45	1,400	671.5
1986	8	47	1,510	781.5
1987	8	47	1,675	781.5
1988	8	47	1,895	891.5
Total	8	528	1,895	891.5

Source: "Ten-Year Energy Development Program, 1979-1988" (Makati, Metro Manila: Ministry of Energy, 1979), p. 42.

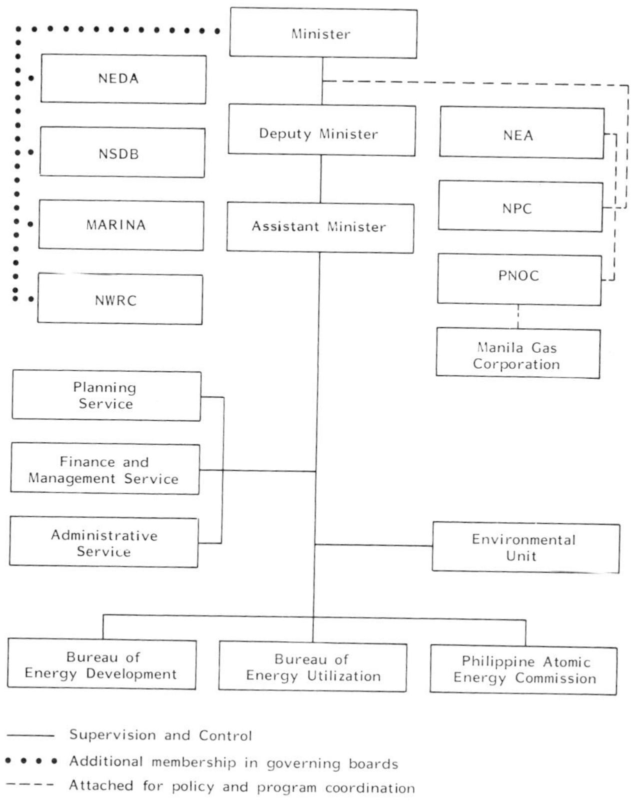

MARINA - Maritime Industry Authority
NEA - National Electrification Administration
NEDA - National Economic and Development Authority
NPC - National Power Corporation
NSDB - National Science Development Board
NWRC - National Water Resources Council
PNOC - Philippine National Oil Company

Fig. 4.6. Ministry of Energy – organizational chart.
Source: "Ten-Year Energy Program, 1979-1988" (Makati, Metro Manila: Ministry of Energy, 1979), p. 99.

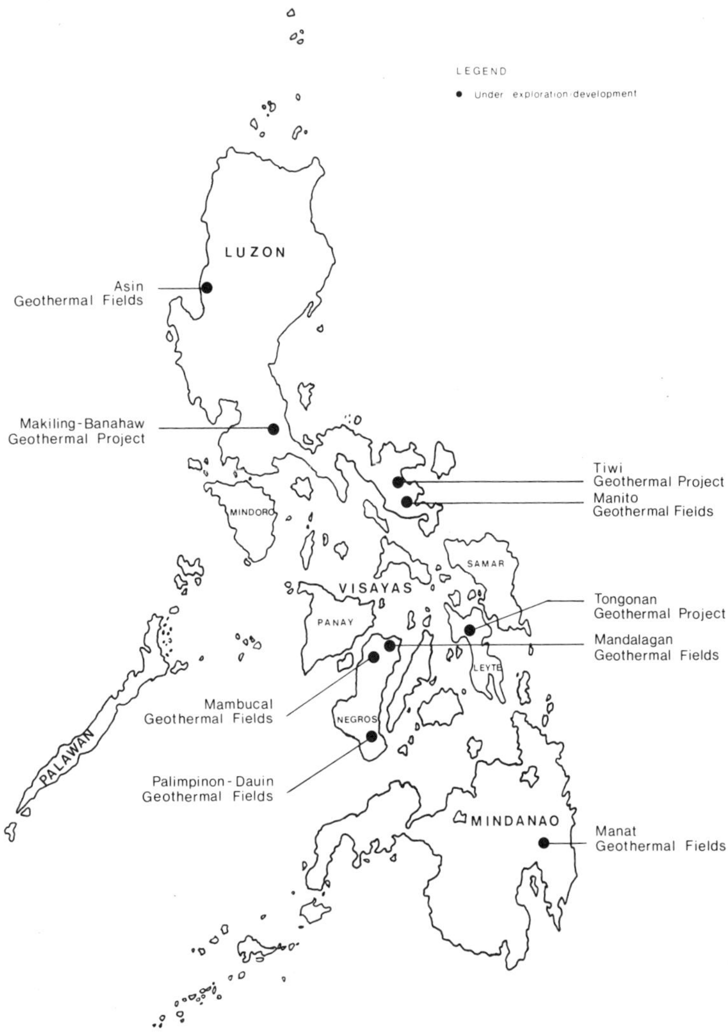

Fig. 4.7. Philippine geothermal fields under exploration or development. Source: "Ten Year Energy Program, 1979-1988" (Makati, Metro Manila: Ministry of Energy, 1979).

program relies to a great extent on the successful planning and implementation of projects, hence the need for effective project managers trained in project management skills and techniques.

The Philippine energy program will still adapt to the changing energy scenario. The continuing validity of the planning effort, as stated in the preface to the "Ten-Year Energy Program, 1979-1988," lies in a regular fine tuning of the entire plan relative to changes in critical decision parameters and constraints. The motive and underlying objectives, however, remain constant. In the words of Energy Minister Geronimo Z. Velasco, the motivation for the entire program is "to provide the consumer with cost-effective energy alternatives that are essential to the economic, social and political progress of our nation."(5)

NOTES

(1) Service contract between Philippine Geothermal, Inc. and the National Power Corporation, September 10, 1971.

(2) Philippines, Office of the President, Presidential Decree No. 40, 1972, "Establishing Basic Policies for the Electric Power Industry," in F.D. Pinpin, ed. The First 107 Presidential Decrees (Mandaluyong Rizal Philippines: Cacho Hermanos, 1973), p. 98.

(3) Philippines, Congress of the Philippines, Republic Act No. 6395, 1971, "An Act Revising the Charter of the National Power Corporation," in Republic Acts, Concurrent Resolutions, and Resolution, 1971 (Makati, Manila: House of Representatives, 1971).

(4) "Ten-Year Energy Development Program 1978-1987" (Makati, Metro Manila: Ministry of Energy, 1978), p. 7.

(5) "Ten-Year Energy Development Program 1979-1988" (Makati, Metro Manila: Ministry of Energy, 1979), preface.

5 Policies of Geothermal Development

Louis J. Goodman
Ralph N. Love

THE CASE HISTORY RECORD

Geothermal energy's role in the future of the world energy scene is uncertain. One authority suggests that "earth heat could one day supplant fossil fuels as the world's main source of energy,"(1) while another considers it unlikely that geothermal energy will ever supply more than 10 percent of the energy needs of the United States.(2) Countries such as the Philippines and New Zealand already place a great deal of importance on the development of geothermal resources; it is estimated that over 10 percent of the total electricity generated in these countries will come from geothermal energy within the next ten years. In contrast, the United States has major geothermal resources, but their development is not seen as a major solution to the country's overall energy problems.

Whatever the future use of geothermal energy may be in these and other countries, the ability to use and develop geothermal resources for generating electricity has been amply demonstrated, and there are few situations in which lack of technical knowledge has precluded the use of geothermal energy resources. New and exciting techniques are constantly being developed to allow greater exploitation of the earth's heat for energy, but technology and its availability are not the main constraint on geothermal development in many countries. In our haste to achieve technical understanding, we have sometimes underestimated the difficulties that accompany the implementation of new types of technology. Indeed, experience demonstrates that the fullest use of what is now recognized as an important energy source is highly dependent upon the ways in which projects for developing geothermal energy are planned and executed. The three cases in this book are vivid illustrations of the fact that resolving the policy issues and implementing the projects create common problems that cut across national boundaries.

Many of the serious problems the case histories reveal, it should be noted, arise from such issues as the legal status of geothermal resources and the type of organizational channels through which this relatively new source of energy can be developed most effectively. Each case presents important lessons in developing policies, methods, and systems to utilize geothermal resources.

The three countries from which the case histories are drawn – the United States, the Philippines, and New Zealand – approach their energy problems through different policy-making procedures and government structures. The strategies each country adopts to cope with its shortfall in energy have in common the fact that they are being determined to a large degree by decisions at the highest level of government. In the United States, however, private-sector utilities are heavily involved in the energy-providing field, creating a different approach to energy development problems from New Zealand's, which has centralized, through its parliamentary government, the ownership and control of electricity-producing energy sources. No private utilities are involved in electricity development or distribution in New Zealand. The Philippines, with its martial law administration, receives direction from the central government, under whose auspices strategies, policies, and priorities are developed. Thus, each of these countries uses a policy decision-making apparatus that operates in a different way to cope with the complex problem of developing energy alternatives.

Wairakei

Wairakei was in many ways a pilot project from which a number of issues emerged which were later used as the basis for refining methods of administering geothermal development projects in New Zealand.

The Wairakei project was, for all practical purposes, the first to utilize a wet geothermal field on a major scale. A dry steam field had been established at Larderello in Italy, but the conditions there were radically different from those at Wairakei. Many of the decisions made had to be based on technical information which was new and, in many cases, untested. Many of the calculations in reaching decisions were hypothetical and judgments could not be made on experience with similar projects. A shortage of coal and other fuel after World War II meant that energy shortages were becoming a major restriction on the development of the country. Consequently, many of the decisions on energy development had to be made on the basis of faith in a situation which demanded prompt action. Thus, this case illustrates that with new technological development, project decisions may have to be made and put into practice even though all the research and development has not been completed.

In the early stages, Wairakei moved from a purely electric power plant to a joint development project between two countries, each with different objectives in mind. First, the New Zealand government wanted to have an alternative electric power source. Second, the

United Kingdom Atomic Energy Administration (UKAEA) was interested in producing heavy water and recognized the opportunity of using the geothermal project for this purpose. A 1955 estimate on the cost of heavy water, however, doubled earlier estimates. Even though the heavy water project had been a major factor in the design phase, it was decided that when the UKAEA pulled out of the project because of increased costs, the power-generating side of the project would continue. This meant that halfway through the project, the New Zealand government had to adapt both the equipment design and administrative facilities to continue the project on a radically different basis from the original intent.

The success of this project and its leadership in the use of geothermal resources is due to several factors. The high priority placed on the project by the government of New Zealand and the appointment of a geothermal advisory committee enabled progress to be made at the highest level, so that all matters requiring attention could be satisfied by one senior interdepartmental government committee. With a new resource, there were many uncertainties about whose responsibility the geothermal resources were. The formation of an interdepartmental government committee responsible directly to the minister meant that it had enough authority and enough contacts in the various government departments to ensure that there were no major delays in the decision-making process.

The creation of Geothermal Development, Limited, a limited-liability company formed from members of the British and the New Zealand governments, was an example of an organizational technique employed as required in other countries, but not that frequently used in New Zealand. It does offer, however, many possibilities, because it provides a ready-made administrative vehicle to undertake the work. The fact that the New Zealand government made a decision to join with the United Kingdom government in the manufacture of heavy water was in fact an important one in reaching a decision to go ahead with the development. At the early stages, it appeared as if the heavy water project would be profitable and would take some of the risk out of the geothermal development. The policy decisions relating to the Wairakei case were mainly centered around the ability to organize and coordinate the various departments and organizations contributing to the development of the project.

Tiwi

The Tiwi case highlighted several important issues regarding a government's energy policies and its approach to implementing geothermal development.

In the Philippines, the need to develop indigenous energy resources was recognized before the oil crisis of 1973. This was significant in light of the fact that the geothermal fields were explored and developed and producing electricity during the 1970s. Geothermal fields that have been explored after the oil crisis may only produce power by the 1980s. This means two things: a longer lead time before the project can be completed when alternative energy is badly needed now, and more expense in terms of development of the wells, since this requires oil fuel.

In the course of the Tiwi project, a number of government entities played a significant role. The relationship of these government agencies and their role in the overall energy program was constantly changing. The initial agencies involved in the project operated independently to a large extent, at times using committees or superstructures to coordinate the work. By the time the project was completed, however, with the inauguration of the first and second power plants, a centralized Ministry of Energy had been created to integrate all energy programs.

Top administrative support for the project ensured its success. Because of the support of the country's president and the assistance of the executive secretary, the project cut through much red tape and was able to proceed at a relatively efficient rate. The president, for example, realized the need to set aside the geothermal field as a government reservation to facilitate the work of exploration and commercial development.

Currently, manpower development in the energy field has been constant in the Philippines and Philippine scientists have been trained both locally and abroad. The government drew on the knowledge and experience of countries in the world that have commercially exploited geothermal energy – the United States, New Zealand, and Italy. Whether consultants are brought into the country, or local scientists go for studies and training in a foreign country, the process of technology transfer has been ongoing. At the start of its geothermal development, the country had to rely on service contractors to develop its fields, but since 1977, Philippine service contractors operating through the Energy Development Corporation have taken over the commercial development of the geothermal fields such as the one in Tongonan, Leyte.

In assuming responsibility for the commercial development of the Tiwi geothermal field, the National Power Corporation (NPC), proved its ability to orchestrate the tremendous job of constructing a power plant. The operations of the NPC are scattered all over the country and cover hydroelectric, nuclear, and geothermal projects. The NPC, too, has been dynamic in its organization and management. Recently, a reorganization was effected, and some projects are now handled on a purely project management basis.

The Tiwi project uncovered several policy issues that also relate to the Hawaii geothermal project. At Tiwi, it was essential to have a single director responsible for the coordination of all phases of geothermal development. The diverse interdisciplinary nature of geothermal projects generally, in fact, requires one person with overall authority. He must mediate among the engineers, geologists, geophysicists, and social scientists; and then he must make key decisions. One suggestion from the reservoir of engineering literature is to train geothermal project managers. They would have a knowledge of geophysics, geology, engineering, environmental sciences, land law, geothermal law, and project management. Most important, such highly trained persons would have a broad perspective and be able most effectively to use project resources in a socially and environmentally acceptable manner.

Hawaii

Geothermal energy affects and therefore involves all segments of society. Some of those affected by a project include the utilities, industrial and home consumers of electricity, landowners, energy policy makers, planners, environmentalists, and those living near the development. Because of this broad spectrum of interest groups, it is necessary to have some sort of policy committee which provides for each group to participate and have some input into the plans for the project. Not only is such participation desirable for efficient implementation of the project, it is also desirable for developing a socially appropriate policy.

The Hawaii geothermal project underscored the confusion in existing land laws in Hawaii. Under existing law, geothermal resources may belong to the state or to the private landowners, depending on whether the court declares them to be mineral, water, or sui generis. If they are declared mineral, they belong to the state because land deeds contain mineral reserve clauses granting ownership of minerals to the state. However, this problem is compounded by the fact that certain territorial land sold to private citizens in the early 1900s did not include the reserve clause. If they are declared to be water, the geothermal resources belong to the private land holders by precedent. If they are classified as sui generis – that is, unique in nature – then a whole new set of laws will have to be adopted. Currently, the law passed by the Hawaii State Legislature defines geothermal resources as mineral. However, this law will be challenged in court.

Further complicating the issue is native rights. The constitution of the state of Hawaii provides that any indigenous natural resource that is exploited for a public purpose must benefit native Hawaiians. Native Hawaiians feel they have a legitimate claim to any geothermal resources. Thus far, no plans have been made to satisfy this claim.

Geothermal plants have to be built near the source of energy. In Hawaii, the source of energy is on the island of Hawaii, which is separated from Oahu, the major industrial and population center, by more than 100 miles of ocean. Under present technology, it is not feasible to transfer geothermal energy to Oahu, where it is most needed. This circumstance raises several issues. First, the reservoir is estimated to be 500 and perhaps even 1,000 MW. This huge source of energy is already attracting major oil corporations and energy-intensive industries, such as processing plants. The government of the island of Hawaii is interested in controlling its growth and development. If its energy potential attracts many large-scale users, government officials are concerned that the overall needs of the state and of private industry will dictate development policies for the island. This raises many implications concerning the control of this resource.

Because the utilities in Hawaii are private, legal monopolies, they have relatively little interest in risking capital in alternative energy development. In essence, the utilities do not have to take risks because their market and their profit are guaranteed. There is no incentive for them to risk development capital. Given this situation, the research and development costs for geothermal research must be underwritten by the government. There is some justification for this policy in the fact that government can distribute the risk over a larger population than can a utility. Moreover, the benefits of geothermal development would accrue to the entire population.

Government involvement extends beyond research and development. Even in the initial steps of utilization, the utilities have expressed a reluctance to be involved until there is absolute proof that they will make a profit. In this instance, there is some justification for a government subsidy, especially since a precedent has already been set by the subsidy for the nuclear industry.

In Hawaii, the role of government in geothermal development extends well beyond the economic sphere. Federal, state, and local governments are responsible for rules and regulations concerning health, safety, transportation, water rights, ownership, occupational safety, energy development, environmental protection, and land use. The Hawaii Geothermal Project clearly showed that geothermal regulations related to these areas would be necessary for future development. Moreover, these regulations would have to be rationalized into a single consistent set of guidelines. The guidelines themselves would have to be well balanced, not only promoting Hawaii's overall needs for energy, but also adequately safeguarding the interests of all groups affected by the development. In sum, it would be the responsibility of government to provide a socially appropriate institutional framework for geothermal development.

THE BROAD ISSUES

A detailed analysis of the history of each of the three projects indicates it is possible to identify six broad issues underlying geothermal development. These are interrelated and may be categorized in terms of planning, policy, and administrative/managerial issues. The six broad issues, which have both macro and micro perspectives, are: 1) present production, current capacity, and future potential; 2) scientific and technical issues; 3) comparative economics; 4) environmental issues; 5) sociopolitical issues; and 6) administrative/managerial issues.

Production, Current Capacity, and Future Potential

At present, geothermal production and capacity in global terms is insignificant. In 1976, geothermal-generated electricity contributed only 0.13 percent to the world's total electrical production; of the world's total installed electrical capacity, only 0.08 percent was geothermal. In the case of the United States, the leading producer of geothermal electricity, only one-tenth of 1 percent of the total electrical production came from geothermal sources. In these terms, geothermal energy has had, as yet, no global impact.

Several factors, however, make geothermal energy much more important than these numbers indicate. First, on an international level, there are over 80 nations actively involved in geothermal research and development. The size and scope of each research program are expanding rapidly. Second, in some developing countries, where the percentage of electricity produced is a small fraction of that produced in the United States, geothermal energy is vitally important.

Today, most countries in the world are faced with a crisis in energy availability. When a similar situation faced New Zealand following World War II, a calculated risk was taken to develop geothermal resources, at that time an unknown and untried energy source. This decision has enabled the industry to develop to a point where geothermal power stations now feed electricity into the national grid and supply up to 11 percent of the total energy needs of the North Island of New Zealand.

In the United States, the geothermal paramarginal reserves – those reserves recoverable at one to two times the cost of present technology – are estimated at 38,000 MW for 50 years. Some estimates place probable reserves at 2,000,000 MW for 50 years. In general, as more surveys are conducted, a greater number of reserves is discovered and the estimates rise.

In El Salvador, 50 percent of the installed electrical capacity is generated from geothermal plants; in the Philippines, the percentage is expected to be 10 percent by 1985; and in New Zealand, the percentage is 9 percent with a potential of 25 percent by 2000. Thus, geothermal energy promises to be a significant factor in terms of international development.

The potential generating capacity of geothermal energy is virtually unlimited. Theoretically, the earth contains sufficient heat to generate enough electricity to satisfy the entire world's demand for the next several hundred years. Moreover, since the source of geothermal heat consists of the earth's entire mantle deep within its interior, geothermal resources can, in principle, exist at any point on earth. The limitations to exploiting the resource include problems associated with government policies and the legal framework for exploiting the geothermal resource, economic considerations, and the development and adoption of techniques to utilize the more complex geothermal resources such as the hot dry rock areas.

In summary, geothermal power has the potential to become a major provider of the future energy needs of mankind. This potential, however, has yet to be fully realized.

Scientific and Technical Issues

The scientific and technical problems associated with geothermal energy development exist because of the predictive uncertainties of geophysical surveys, a lack of empirical data on the earth's geologic structure and the processes that occur within the earth, the relatively low efficiency of converting geothermal energy into electricity, and the limited ability to utilize either magma or hot dry rock, which constitute the greatest proportion of the geothermal resource. Research is being conducted in all of these areas, and as the demand for new sources of energy increases, greater resources are being used in research and development programs for utilizing geothermal energy.

The development of small geothermal power plants in the 0.5-to 5-MW range has opened up opportunities for using geothermal energy in many rural and underdeveloped areas. Small geothermal power plants have distinct advantages over other generating methods in these locations; fuel does not have to be transported and transmission costs can be minimized. An international research effort to develop small power plants was started in October, 1973, with France, Italy, Mexico, New Zealand, Portugal, Turkey, and the United States actively engaged in research work in this area. Small power plants presently in existence or in the construction or planning stage are given in table 5.1.

The cooperative research being undertaken will increase the versatility of small power plants and improve the overall economics of using small-scale geothermal power.(3)

Research and development have also been undertaken by many countries in the important field of the direct (nonelectric) applications of geothermal energy. The direct use of geothermal resources is a more efficient use of the resource than converting it to electrical energy.

International research is also addressing the important problems of reservoir mechanics, the identification and assessment of geothermal fields, drilling and production techniques, problems of scaling and corrosion, and methods of minimizing environmental problems.

The research and technical developments which could have the most impact on the future development of geothermal resources are those now being undertaken on hot dry rock. Hot dry rock consists of geological formations below the surface of the earth which have a very high heat content but not sufficient water or are not sufficiently permeable to allow the withdrawal of heat by water or steam. Two important problems are being considered by scientists. First, although it is generally accepted that there are large amounts of heat below the surface, it is difficult to find the resource and assess its potential. Second, is the problem of extracting the resource in an economic manner; for this to happen, new techniques and new equipment in fields

Table 5.1. Small Power Plants Presently Existing, under Construction, or Planned in Participating Countries

Countries	Geothermal Field	Type of Power Plant	Capacity (MW)	Output (MW)	On Line
Japan	Kyushu, Otake	Flash, condensing	13.0 MW	10.0 MW	1967
	Onuma	Flash, condensing	12.5 MW	10.0 MW	1973
	Kyushu, Otake	Binary	1.0 MW	1.0 MW	1977
Philippines	Leyte	Flash, noncondensing	3.0 MW	3.0 MW	1977
	Los Banos	Flash, noncondensing	1.3 MW	1.2 MW	1977
Portugal	Sao Miguel	Flash, condensing	5.0 MW		1979
Turkey	Kizildere	Flash, noncondensing	0.5 MW	0.5 MW	1976
El Salvador	Ahuachapan	Flash, noncondensing	1.3 MW	1.1 MW	1975
United States	Roosevelt	Helical screw	1.2 MW	1.2 MW	1978
	Ralph River	Binary	5.0 MW		1979
	Puna, Hawaii	Flash, condensing	3.0 MW		1979

Source: J.C. Bresee, W.W.S. Yen, and J.E. Metzler, eds., Geothermal Pilot Study Final Report: Creating an International Geothermal Energy Community (NATO Committee on the Challenges of Modern Society, June 1978), p. 42.

such as drilling, fracturing, and mapping are being developed. In the United States, a hot dry rock program began in 1972 with two main goals. First, it seeks to establish the feasibility of the hydraulically fractured hot dry rock system.(4) Second, it wants to confirm the technical feasibility and to obtain economic data for commercial-size dry rock extraction systems by 1985.

Comparative Economics

From a historical perspective, geothermal plants have been significantly more cost-effective than many fossil fuel and nuclear plants. The 1961-1974 average generating cost for The Geysers was 5.6 mills/kW compared to the local average of 8.5 mills/kW for nuclear plants. Geothermal costs in other parts of the world are similar. The cost at Larderello, Italy has been 3.2 mills/kW and in Matsukawa, Japan, 4.6 mills/kW. When installed capital costs are included in the overall cost, geothermal energy production becomes even more advantageous. The value of historical comparisons is negligible, however, because the cost of power generation is site and plant specific. Therefore, it is more useful to compare the projected costs of alternatives at a specific site.

Projecting the economic costs of alternative sources of power generation is an extremely uncertain task because projected costs must be based upon a wide variety of underlying assumptions about the future. Assumptions must be made about numerous variables such as resource availability, exploration costs, interest rates, labor costs, drilling depths, the available technologies, and fuel costs. Additionally, cost estimates for geothermal plants fluctuate greatly, depending upon the quality and composition of the geothermal source. Some of the uncertainty can be reduced when projecting the costs for specific plants at specific sites. In projecting any generalized cost comparisons of competing sources of energy, however, much of the uncertainty remains. Given these notes of caution, most theoretical models of comparative costs show geothermal to be competitive with other sources of power generation (see table 5.2).

Environmental Issues

Environmental issues depend largely on the specific geothermal site. Certain environmental concerns, however, cut across the entire range of geothermal developments. These include chemical discharges, noise, heat discharges, subsidence, and chemistry. Of the chemical discharges, hydrogen sulfide is the most widely documented: at The Geysers, emissions are estimated to be 28 tons/day (25,401 kilograms/day); at Wairakei estimates are 40 tons/day (36,287 kilograms/day); and at Cerro Prieto, Italy, estimates are 55 tons/day (49,895 kilograms/day). Other chemical discharges that cause concern are carbon dioxide, which is not yet a pressing problem but will have to be addressed in the future, silica, which has resulted in the destruction of small areas of coffee plants in El Salvador, and salt (NaCl) which has destroyed some shrubs and similar vegetation at Wairakei. Additionally, at Wairakei the geothermal effluent is released into the Waikato River, and subsequent chemical analyses have indicated traces of boron, mercury, and arsenic in the water. A related problem at Wairakei is the heat of the discharge, which resulted in an average $2^{o}C$ ($3.6^{o}F$) increase in the water temperature and at times leads to a $6^{o}C$ ($10.8^{o}F$) rise. The temperature increase has resulted in interference with trout hatching and an increase in the number of fish killed, although impact upon the fish is localized to the area near the effluent discharge.

The only area which reports significant subsidence is Wairakei, parts of which have subsided as much as 10 meters (33 feet) since 1950. An unknown environmental hazard is the threat of increased seismicity from reinjection. In Colorado, some experiments demonstrated that reinjection could cause severe shocks, but in Italy a carefully controlled experiment revealed no seismic effects from reinjection. More information is needed on the impact of reinjection of geothermal fluids.

Table 5.2. Comparison of Fossil-Fuel, Nuclear, and Geothermal Estimated Generating Costs (in 1976 Dollars)

Resource Type	Installed Equipment Costs ($/kW)	Equipment Cost as (cent/kWh)[a]	Operating and Maintenance as (cent/kWh)	Well or Fuel Cost (cent/kWh)	Total Generating Cost (cent/kWh)
Direct flashing[b]	300-600	0.68-1.37	0.13	0.80-2.80	1.61-4.30
Binary-fluid cycles[b]	400-700	0.91-1.60	0.13	0.53-2.45	1.57-4.18
Nuclear[d]	$\geq$800	$\geq$1.83	0.13	0.30	$\geq$2.26
Fossil fuel – oil[e]	400-600[c]	0.91-1.37	0.13	2.0($12/bbl)	3.04-3.50
Fossil fuel – cost[e]	400-600[c]	0.91-1.37	0.13	1.0($25/ton)	2.04-2.50

[a] 171 annual fixed charge rate. 851 (7,446 hr/yr) load factor.
[b] 150-200°C resources. Plant startup in 1980.
[c] Higher costs correspond to stringent environmental control systems.
[d] Plant startup in 1984.
[e] Plant startup in 1980.

Source: Reprinted from Geothermal Energy as a Source of Electrical Power by Stanley Milora and Jefferson Tester, by permission of The MIT Press, Cambridge, Massachusetts. Copyright © 1976 by The Massachusetts Institute of Technology.

Sociopolitical Considerations

Many experts believe that one of the major obstacles to developing geothermal resources in the United States is the legal framework. The complex issues of ownership and regulation are often ambiguous.

Moreover, there is no single comprehensive legal framework to guide geothermal development because the federal government, state governments, and local governments each have their own laws and regulations regarding geothermal exploitation. Some of these issues have been addressed and resolved by other nations. In New Zealand, geothermal resources are reserved to the Crown, that is, they are owned by the state, regardless of who holds title to the land. A single framework of rules and regulations is provided by the government regarding exploration and development. These include safety regulations and environmental regulations that set forth specific standards which the developers must meet. In Iceland, geothermal resources belong to the owners of the surface property, and the right to utilize the resource is subject only to regulations set by the state. One of these regulations, however, is the requirement to obtain a concession from the parliament to develop commercial-scale geothermal energy. In practice, only the publicly owned utilities have obtained such concessions.

One issue not sufficiently addressed by most nations is public participation in geothermal policy making. As demonstrated by recent events in the energy field, it is necessary to develop some institutionalized process for public participation in policy making. With geothermal resources, this is particularly appropriate to local areas, since first-order geothermal impacts are localized.

Even more important than the legal framework is the institutional framework. In every country where utilization of geothermal resources has been vigorously pursued, there has been a centralized public agency responsible for energy development. In New Zealand, although many ministries are partly responsible for energy rules and regulations, the Ministry of Works and Development has been responsible for geothermal development and implementation. In the Philippines, the National Power Corporation, under mandate from the central government, has been responsible for the active program in geothermal development. In Iceland, the publicly owned utilities have developed the geothermal resource. In El Salvador, vigorous geothermal development has been conducted under the auspices of the national electric company. In each of these instances, the costs of geothermal energy implementation have been distributed over the entire population because central government agencies are responsible for energy development. This circumstance allowed geothermal energy utilization to be pursued within the framework of national balance of payments, national resources, and long-term national goals, with less regard to achieving a high rate of return on the investment.

What have been some of the social and political impacts of geothermal energy development? In New Zealand, the Wairakei project has developed as a tourist attraction, with over 100,000 visitors a year. This has stimulated a local tourist industry which has contributed to the employment opportunities for Maoris living in the area. The geothermal development has a permanent work force of 250, and utilizes land that otherwise would probably have been used for timber or pastoral farming.

In the Philippines, the geothermal plants will be major sources of electrical production for the nation and are expected to contribute greatly to the development effort. Since geothermal plants must be located where the resources exist, however, urban industrial areas and regional development centers are being planned and developed in rural areas which might not have otherwise been developed.

In Italy, the government is pursuing a policy of electrifying the rural areas by means of small-scale geothermal plants. Geothermal energy has thus allowed Italy to adopt a relatively decentralized and localized approach to modernization, avoiding the dislocations of a crash program.

Administrative/Managerial Issues

Administrative and managerial issues encompass four separate but interrelated components: 1) institutional structure; 2) international organization; 3) personnel; and 4) the management plan. It is apparent in each of the case histories that the ideal situation calls for top government administrative support for the project, ranging from viable policy guidelines to availability of necessary resources – manpower skills, capital, technology, equipment, and materials.

In each of the projects, the development of an appropriate institutional structure had to be considered, as the development of geothermal resources was a new experience for the organizations concerned. The need for close cooperation among a number of government and private organizations, each with a vital contribution to make to the project, meant new organizational and management structures had to be developed to cope with the needs of the project.

Each of the cases illustrates the difficulties faced in developing an appropriate institutional structure and an efficient organization to control the project. Thus, the work at Wairakei called for a particularly close integration of roles of several New Zealand government departments, creating in effect a new type of working relationship. Determining who had responsibility for certain key areas of study and implementation initially created some difficulties.

A major feature of the managerial framework during the Wairakei project's initial stages was the formation of the Geothermal Advisory Committee. The GAC was established by authority of the government in 1949 from amongst existing government organizations. This meant that the GAC reported directly to the government as a group rather than reporting individually by department to the government through its own ministry. This coordination of the three major departments involved with the Wairakei development meant that when fast decisions needed to be made, they could be made by the GAC with the full authority and support of the departments involved. Unnecessary delays caused by fragmented decision making could consequently be avoided. For example, in 1952 the New Zealand government, in the face of an energy crisis and an election year, decided that the Wairakei group had to prove in 12 months the existence of 20,000-kW geothermal steam. The investigation program had already been condensed from what was regarded as a tight program and now the government was pressing harder to get early results. The GAC, because of its authority by virtue of its appointment by the government and its consolidation of the departments involved, was able to act quickly and effectively to meet the new 12-month deadline. The GAC set up a subcommittee consisting of an officer each from MWD, SHED, and the geological survey office and asked them to come up with a plan. The plan was to drill enough holes to produce 20,000 kW and let the holes discharge for a period of time.

Another major feature of the case was the proposal to produce both heavy water and power from the geothermal resource and the consequent problems of the joint effort of the New Zealand and the United Kingdom governments. Wairakei would have been developed completely differently if there had never been the possibility of producing heavy water.

The difficulties surrounding the project when there was to be joint development for power and heavy water centered on problems of coordination. There was a need to coordinate the activities of a number of groups working in different places, each with somewhat different views of the same objective. A symptom of the coordination problem was that of the conflicts over estimates. When the first estimates came in from the consultants, one estimate by New Zealand "power only" consultants and the other by United Kingdom Atomic Energy Authority "heavy water plus power" consultants, there were fundamental differences between the two groups and their estimates because of the two different approaches they took. The Commissioner of Works, who was in charge of the project, suggested the two groups of consultants get together either to reconcile their reports or to explain the reasons why they could not. It was clear that more information was needed for the first estimates; when more information became available, the substantial increase between the 1954 and 1955 heavy water estimates meant decision makers had to drastically revise their thinking.

Providing estimates based on inadequate information can lead to major problems for decision makers. As it turned out, the heavy water advocates pulled out of the project because of the cost involved, emphasizing the organizational problems that can arise in a project the nature of whose work continually changes from scientific investigation to engineering development. But the conflicts that arose illustrate the need to establish clear managerial authority and procedures, and systems for conflict resolution when two such groups with different approaches are working on one project.

In the Hawaii case, an interesting management sidelight was the handling, by the project director and his executive committee, of a situation within the administrative organization that could have caused major problems. The geophysics program director steadfastly believed, in opposition to the general feeling among the other decision makers, that further geophysical surveys should be conducted before the initial drilling took place. He had also disagreed with the site selection committee's decision and had urged them to drill in Area B rather than Area A. Thus, when the testing period began in July 1976, the project director and the executive committee felt that the geophysics program director's continuing differences of opinion about project direction and policy made it difficult for him to work effectively with the entire Hawaii Geothermal Project (HGP) team. This, plus the fact that they wanted to broaden the scope of the scientific investigation by integrating the geological, geophysical, and geochemical components of the project into one "geoscience" program, prompted them to take action. They formed the new program, and named a new director as its head.

The biggest environmental problem that the HGP program encountered was local residents' strong objections to the "bloodcurdling banshee howl" that pierced the normally quiet air in that serene corner of the island of Hawaii during the testing period. This problem was dealt with by maintaining close ties with the community, especially through the project director and the well-head testing project manager who was a resident of the island.

The successful development of geothermal energy in Hawaii was aided in no small degree by the project director's efforts in forming the Hawaii Advisory Committee (HAC) and the National Liaison Board (NLB). The HAC was composed of leaders such as the president of Hawaii's electric company, the director of the state Department of Planning and Economic Development, the director of a leading environmental group, the president of the Congress of Hawaiian People, and officials from the county of Hawaii. These people represented those groups that formulated Hawaii's energy policy, so their support was critical. The NLB was composed of geologists, geophysicists, and engineers – experts in geothermal power development. Besides representing a core of the nation's geothermal experts, they worked for key agencies such as the National Science Foundation and the United States Geological Survey. Thus, they were very influential in ensuring federal funding for the project. The importance of HAC and NLB to the success of the HGP illustrates the necessity of enlisting the support of key figures as far as the policy and funding of a project are concerned.

Perhaps the dominant feature of the success of the Philippine Tiwi project, from the managerial point of view, was the top administrative support it received. With the authority of the country's president and executive secretary behind it, the project was able easily to clear hurdles that might have caused problems in other countries. For example, nearly 18,000 hectares (44,478 acres) were set aside for exploration and development of the geothermal resource. Also, as a corporation owned wholly by the Philippine government, the NPC had support at the highest levels of government. As a country whose dual policy it was to reduce its dependence on foreign oil and strike a balance between urban and rural development by harnessing indigenous energy to increase power in rural areas, the Philippines found little or no opposition to the development of its geothermal resources in the case of Tiwi.

CONCLUSION

The growing importance of geothermal energy and the need for specialized manpower skills to develop geothermal resources is now being recognized. For example, the United Nations Development Program in conjunction with the New Zealand government has established at the University of Auckland, New Zealand a geothermal institute which offers a postgraduate diploma in geothermal energy technology. The course, designed to cater to participants from all developing countries, is aimed particularly at earth scientists (geologists, geochemists, geophysicists) and engineers (chemical, civil, mechanical) who can utilize their basic skills in developing geothermal resources. The need for skilled manpower trained in the development and utilization of geothermal resources is now accepted as a key ingredient to the future successful development of the earth's heat.

To take the broadest view, the three case histories presented here illustrate that the management of geothermal projects must be capable of the following tasks: 1) coordinating the resource exploration and evaluation activities and establishing optimum plant location; 2) establishing the size of the resource and its hydrothermal and chemical characteristics; 3) designing the site's civil improvements and plant module foundations; 5) designing the collection and reinjection system pipelines; 6) managing the procurement and detailed design of all modules; 7) designing and procuring the module interconnections; 8) specifying and procuring appropriate alternative or optional units; and 9) managing the construction, activation, and training activities associated with fully implementing a geothermal power plant.(5)

The method by which these nine activities were achieved differed in the three projects. Each project, however, had to utilize the same technical and managerial skills to bring their operations to the point where power was actually produced. Future geothermal projects will follow the same path, but the advances in technical and managerial skills will be available to ensure project objectives are met in the most effective way.

Beyond these considerations, the cases demonstrate that geothermal power development involves numerous separate but interrelated tasks. These tasks are accomplished by the combined actions of government policy makers, government departments, private organizations, and specialist consultants. Working together, they must establish firm legal guidelines for utilizing and developing the geothermal resource and provide the organization, manpower, and other materials necessary to

DEVELOP THE PROJECT.

develop the project. The final lesson of the case histories, in fact, is that the actual development of a geothermal project depends almost entirely on the complex policy and decision-making environment that surrounds it. For this policy environment, after all, will ultimately determine whether or not geothermal energy enters the mainstream of world energy consumption.

NOTES

(1) H. Christopher H. Armstead, Geothermal Energy (London: E. and F. Spon, 1978), p. 32.

(2) J.D. Worthington, "Geothermal Development," in Status Report: Energy Resources and Technology Report of the Ad Hoc Committee on Energy Resources and Technology, Atomic Industrial Form, Inc., 1975.

(3) J.C. Bresee, W.W.S. Yen, and J.E. Metzler Geothermal Pilot Study Final Report: Creating an International Geothermal Energy Community (Berkeley: University of California Press: June 1978).

(4) Ibid.

(5) NATO Committee on the Challenges of Modern Society, Geothermal Pilot Study Final Report: Creating an International Geothermal Energy Community. June 1978, pp. 35-36.

Index

About the Authors

GENERAL EDITORS AND CONTRIBUTING AUTHORS

LOUIS J. GOODMAN joined the East-West Center Staff in 1971 after serving as a Project Specialist in Engineering Education with the Ford Foundation in the Philippines from 1968-1971. He is Assistant Director of the Resource Systems Institute. Goodman has twenty-two years of national and international experience as an engineering educator, researcher and consultant. He holds degrees from M.I.T. and Harvard University and is registered as a professional engineer to practice in the states of Ohio and New York.

RALPH NGATATA LOVE is currently Director of the Management Education and Development Centre, Massey University, Palmerston North, New Zealand. Before taking up an academic post in 1970, he worked in industry for fifteen years and now serves as director of several private and public corporations. Born in New Zealand, he studied at Victoria University of Wellington and holds degrees in Commerce and Business Administration and a Ph.D. in Policy Formation.

CONTRIBUTING AUTHORS

ROSEMARY M. AQUINO, Adjunct Professor and Director, Project Management Center, De La Salle University, Manila, Philippines.

SALVADOR P. AQUINO, Manager, Manpower Development, Philippine National Oil Company, Manila, Philippines.

RICHARD S. BOLTON, Chief Geothermal Engineer, Ministry of Works and Development, New Zealand.

TETSUO MIYABARA, East-West Center, Honolulu, Hawaii, Research Fellow, Resource Systems Institute.

BARBARA YOUNT, East-West Center, Honolulu, Hawaii, Writer/Editor, Resource Systems Institute.

Other Contributors

ARTURO ALCARAZ, Geothermal Consultant, Philippine National Oil Company, Energy Development Corporation, Ministry of Energy, Philippines.

JOHN SHUPE, Dean of the College of Engineering, University of Hawaii, and Director, Hawaii Geothermal Project, United States.